LONDON MATHEMATICAL SOCIETY STUDENT TEXTS

Managing Editor: Ian J. Leary,

Mathematical Sciences, University of Southampton, UK

64 A short course on Banach space theory, N. L. CAROTHERS
65 Elements of the representation theory of associative algebras I, IBRAHIM ASSEM, DANIEL SIMSON & ANDRZEJ SKOWROŃSKI
66 An introduction to sieve methods and their applications, ALINA CARMEN COJOCARU & M. RAM MURTY
67 Elliptic functions, J. V. ARMITAGE & W. F. EBERLEIN
68 Hyperbolic geometry from a local viewpoint, LINDA KEEN & NIKOLA LAKIC
69 Lectures on Kähler geometry, ANDREI MOROIANU
70 Dependence logic, JOUKU VÄÄNÄNEN
71 Elements of the representation theory of associative algebras II, DANIEL SIMSON & ANDRZEJ SKOWROŃSKI
72 Elements of the representation theory of associative algebras III, DANIEL SIMSON & ANDRZEJ SKOWROŃSKI
73 Groups, graphs and trees, JOHN MEIER
74 Representation theorems in Hardy spaces, JAVAD MASHREGHI
75 An introduction to the theory of graph spectra, DRAGOŠ CVETKOVIĆ, PETER ROWLINSON & SLOBODAN SIMIĆ
76 Number theory in the spirit of Liouville, KENNETH S. WILLIAMS
77 Lectures on profinite topics in group theory, BENJAMIN KLOPSCH, NIKOLAY NIKOLOV & CHRISTOPHER VOLL
78 Clifford algebras: An introduction, D. J. H. GARLING
79 Introduction to compact Riemann surfaces and dessins d'enfants, ERNESTO GIRONDO & GABINO GONZÁLEZ-DIEZ
80 The Riemann hypothesis for function fields, MACHIEL VAN FRANKENHUIJSEN
81 Number theory, Fourier analysis and geometric discrepancy, GIANCARLO TRAVAGLINI
82 Finite geometry and combinatorial applications, SIMEON BALL
83 The geometry of celestial mechanics, HANSJÖRG GEIGES
84 Random graphs, geometry and asymptotic structure, MICHAEL KRIVELEVICH *et al*
85 Fourier analysis: Part I - Theory, ADRIAN CONSTANTIN
86 Dispersive partial differential equations, M. BURAK ERDOĞAN & NIKOLAOS TZIRAKIS
87 Riemann surfaces and algebraic curves, R. CAVALIERI & E. MILES
88 Groups, languages and automata, DEREK F. HOLT, SARAH REES & CLAAS E. RÖVER
89 Analysis on Polish spaces and an introduction to optimal transportation, D. J. H. GARLING
90 The homotopy theory of $(\infty,1)$-categories, JULIA E. BERGNER
91 The block theory of finite group algebras I, MARKUS LINCKELMANN
92 The block theory of finite group algebras II, MARKUS LINCKELMANN
93 Semigroups of linear operators, DAVID APPLEBAUM
94 Introduction to approximate groups, MATTHEW C. H. TOINTON
95 Representations of finite groups of Lie type (2nd Edition), FRANÇOIS DIGNE & JEAN MICHEL
96 Tensor products of C^*-algebras and operator spaces, GILLES PISIER
97 Topics in cyclic theory, DANIEL G. QUILLEN & GORDON BLOWER
98 Fast track to forcing, MIRNA DŽAMONJA
99 A gentle introduction to homological mirror symmetry, RAF BOCKLANDT
100 The calculus of braids, PATRICK DEHORNOY
101 Classical and discrete functional analysis with measure theory, MARTIN BUNTINAS
102 Notes on Hamiltonian dynamical systems, ANTONIO GIORGILLI
103 A course in stochastic game theory, EILON SOLAN

London Mathematical Society Student Texts 104

Differential and Low-Dimensional Topology

ANDRÁS JUHÁSZ
University of Oxford

Shaftesbury Road, Cambridge CB2 8EA, United Kingdom

One Liberty Plaza, 20th Floor, New York, NY 10006, USA

477 Williamstown Road, Port Melbourne, VIC 3207, Australia

314–321, 3rd Floor, Plot 3, Splendor Forum, Jasola District Centre, New Delhi – 110025, India

103 Penang Road, #05–06/07, Visioncrest Commercial, Singapore 238467

Cambridge University Press is part of Cambridge University Press & Assessment, a department of the University of Cambridge.

We share the University's mission to contribute to society through the pursuit of education, learning and research at the highest international levels of excellence.

www.cambridge.org
Information on this title: www.cambridge.org/9781009220606
DOI: 10.1017/9781009220613

First published 2023

A catalogue record for this publication is available from the British Library.

ISBN 978-1-009-22060-6 Hardback
ISBN 978-1-009-22057-6 Paperback

Contents

Preface

The purpose of this book is to provide a brief introduction to the main concepts and results in differential and low-dimensional topology to an advanced undergraduate or early-stage graduate student. There are several excellent textbooks and research papers in the literature, which we reference, that go into much more detail about most of the topics covered here, and the reader is encouraged to consult them. This book intends to provide a broad foundation and to highlight connections between topics often treated in isolation. We hope that, after reading this short treatise, the reader will be ready to study the most recent tools in low-dimensional topology.

We assume a working knowledge of homology, cohomology, and homotopy theory; see Hatcher [57]. The point of view we take is that of differential topology (i.e., we work in the smooth category) for the most part, though we use techniques from triangulations in the proof of the loop theorem and in our discussion of normal surface theory.

The first two chapters give an overview of classical results on the classification of high-dimensional smooth manifolds, where 'high' means at least five. The main tool is the h-cobordism theorem. This part of the text culminates in the proof of the fact that there are 28 non-diffeomorphic smooth structures on the seven-sphere using surgery theory, due to Kervaire and Milnor [82]. We have tried to keep the discussion here as elementary as possible.

Many of the techniques developed in the first part of the book are used in the second part on low-dimensional manifolds and knots. The exotic phenomena encountered in low dimensions are due to the failure of the h-cobordism theorem. Hence, we believe the high-dimensional theory should be studied before learning low-dimensional topology to get a more complete understanding of the history and context. However, it is possible to skip the first part and only refer to concepts introduced there when referenced from the second part.

Low-dimensional topology can be studied from a number of points of view, and we would like to show a glimpse of each. However, we have done this without the aim of completeness, and our main goal was to lead students to a level where they can start studying three-manifolds and four-manifolds from the point of view of gauge-theoretic and Floer-theoretic invariants. Along the way, we highlight some recent results in each area. We conclude the second part with a brief overview of Heegaard Floer homology, developed by Ozsváth and Szabó [134][138], and the Seiberg–Witten four-manifold invariants [121].

Some of the material in this book forms part of the fourth-year lecture course 'Low-Dimensional Topology and Knot Theory' at the University of Oxford. For an undergraduate audience familiar with homology, cohomology, and the fundamental group, Chapter 1, Sections 2.2 and 2.3 (without the proof of the h-cobordism theorem), Chapter 3 excluding the last two sections, Chapter 4 without the last section, and the first two sections of Chapter 6 could form the basis of an introductory course on the subject. The topics left out are more suitable for a graduate course.

Notation

We do not write the coefficients when considering homology or cohomology over $\mathbb{Z}$. So, for example, $H_i(X)$ denotes $H_i(X;\mathbb{Z})$. Throughout, '$\approx$' stands for 'homeomorphism' or 'diffeomorphism', depending on the context, and '$\cong$' for 'homotopy equivalence' if it relates topological spaces, or 'isomorphism' if it relates algebraic objects. Usually, I denotes the interval $[0,1]$. If S is a subset of a topological space X, then $N(S)$ stands for an open neighbourhood of S. When S is a submanifold of a manifold X, then $N(S)$ is assumed to be an open tubular neighbourhood of S.

Acknowledgements

Firstly, I would like to thank my father, István Juhász, for his encouragement. I am also extremely grateful to David Gabai, Zoltán Szabó, and András Szűcs for giving the wonderful lecture courses that have shaped the directions taken in this book.

Finally, I would like to thank Zsombor Fehér, Allen Hatcher, András Szűcs, and Vladimir Turaev for numerous comments and corrections.

1
Background on Topological and Smooth Manifolds

Manifolds form a particularly nice class of topological spaces that appear in many areas of mathematics. We first give a brief introduction to topological and smooth manifolds, but will mostly focus on smooth manifolds thereafter. We do assume some basic familiarity with one-manifolds and surfaces. We then give an overview of Morse theory, which allows one to study smooth manifolds using the critical points of smooth functions they admit. This leads us to handle decompositions. These can be thought of as CW decompositions where the cells have been thickened to all be of the same dimension. We conclude this chapter with a discussion of cobordism, which provides a coarser notion of equivalence for manifolds than homeomorphism.

1.1 Topological Manifolds

For a non-negative integer n, an n-dimensional manifold (or n-manifold, in short) is a topological space that locally looks like Euclidean space $\mathbb{R}^n$. We now give a rigorous definition.

Definition 1.1 An *n-dimensional (topological) manifold M* is a topological space such that

(i) each point of M has a neighbourhood homeomorphic to $\mathbb{R}^n$,
(ii) M is second countable (i.e., it has a countable basis of open sets), and
(iii) M is Hausdorff (i.e., different points have disjoint neighbourhoods).

Exercise 1.2 Give examples of connected topological spaces that satisfy exactly two of conditions (i)–(iii).

Topological manifolds form a sub-category of the category of topological spaces, where the morphisms are continuous maps. As for topological spaces, isomorphisms are the homeomorphisms.

We now look at some examples. Clearly, $\mathbb{R}^n$ is an n-manifold. So is the *n-sphere*

$$S^n := \{x \in \mathbb{R}^{n+1} : |x| = 1\}.$$

The *real projective space*

$$\mathbb{RP}^n := (\mathbb{R}^{n+1} \setminus \{0\})/\mathbb{R}^* = S^n/\{\pm 1\}$$

is the set of lines in $\mathbb{R}^{n+1}$. The *complex projective space*

$$\mathbb{CP}^n := (\mathbb{C}^{n+1} \setminus \{0\})/\mathbb{C}^* = S^{2n+1}/S^1$$

is the set of complex lines in $\mathbb{C}^{n+1}$ and has (real) dimension $2n$. All the familiar matrix groups such as $\mathrm{GL}(n)$, $\mathrm{O}(n)$, and $\mathrm{SO}(n)$ are manifolds. (In fact, they are Lie groups.)

Manifolds with boundary will also play an important role. We write

$$\mathbb{R}^n_+ := \{(x_1, \dots, x_n) \in \mathbb{R}^n : x_n \geq 0\}.$$

Definition 1.3 An n-dimensional *manifold with boundary* is a topological space M such that

(i) each point of M has a neighbourhood homeomorphic to $\mathbb{R}^n$ or $\mathbb{R}^n_+$,
(ii) M is second countable, and
(iii) M is Hausdorff.

The *interior* $\mathrm{Int}(M)$ of M consists of those points of M that have neighbourhoods homeomorphic to $\mathbb{R}^n$, and its *boundary* is $\partial M := M \setminus \mathrm{Int}(M)$.

Note that, if a point of M has a neighbourhood homeomorphic to $\mathbb{R}^n$, then it also has a neighbourhood homeomorphic to $\mathbb{R}^n_+$, so we cannot define ∂M as the set of points of M having a neighbourhood homeomorphic to $\mathbb{R}^n_+$. The boundary ∂M is a manifold without boundary.

For example, the *n-disk*

$$D^n := \{x \in \mathbb{R}^n : |x| \leq 1\}$$

is a manifold with boundary S^{n-1}. We will write B^n for the open *n-ball* $\mathrm{Int}(D^n)$. Note that every manifold is a manifold with (empty) boundary. We say that a manifold is *closed* if it is compact and has no boundary.

Definition 1.4 A subset N of an m-manifold M is called a *submanifold* if it is a manifold with the subspace topology. We say that N is a *locally flat submanifold of M* if each point $p \in N$ has a neighbourhood $U_p \subset M$ such that the pair $(U_p, U_p \cap N)$ is homeomorphic to $(\mathbb{R}^m, \mathbb{R}^n)$ for some n.

Example 1.5 Let K be a non-trivial knot in S^3; that is, a connected, locally flat one-dimensional submanifold of S^3 such that the pair (S^3, K) is not homeomorphic to (S^3, S^1). Then the cone on K from the centre $\underline{0}$ of D^4 is a submanifold of D^4, but it is not locally flat at $\underline{0}$.

Definition 1.6 A closed, connected n-manifold is called *orientable* if

$$H_n(M) \cong \mathbb{Z}.$$

If M is orientable, an *orientation* of M is a choice of generator of $H_n(M)$. This generator is called a *fundamental class.*

Note that the preceding isomorphism is not canonical, and hence there is no preferred orientation. Our homological definition of orientation captures the intuition of having a coherent system of local orientations in the following sense. Let $\alpha \in H_n(M) \cong \mathbb{Z}$ be a generator. Then, for every point $p \in M$, the image of α under the map

$$H_n(M) \to H_n(M, M \setminus \{p\})$$

induced by the embedding $(M, \varnothing) \to (M, M \setminus \{p\})$ is a generator of

$$H_n(M, M \setminus \{p\}) \cong \mathbb{Z}.$$

More generally, if M is a compact, connected manifold with boundary, then M is orientable when $H_n(M, \partial M) \cong \mathbb{Z}$, and an orientation is a generator of this group. A fundamental example of a non-orientable manifold is the Möbius band. A two-manifold is orientable if and only if it does not contain a Möbius band.

Exercise 1.7 Construct a CW decomposition of $\mathbb{RP}^n$, and use this to compute the homology groups $H_*(\mathbb{RP}^n; \mathbb{Z})$, $H_*(\mathbb{RP}^n; \mathbb{Z}_2)$, and the cohomology ring $H^*(\mathbb{RP}^n; \mathbb{Z}_2)$. For what n is $\mathbb{RP}^n$ orientable?

One of the main problems of manifold topology is the classification of manifolds.

Exercise 1.8 Show that a zero-manifold is a discrete, countable topological space.

We shall focus on connected manifolds, as every other manifold is a countable disjoint union of such. We state the classification of one-manifolds. For a proof, see, for example, Fuks–Rokhlin [42] or Milnor [116].

Theorem 1.9 *Every connected one-manifold with boundary is homeomorphic to one of* S^1, $\mathbb{R}$, $[0, \infty)$, *or* $I := [0, 1]$.

Given manifolds with boundary M and N, we can form their *product* $M \times N$, which has dimension $\dim(M) + \dim(N)$. For example, the *n-torus* T^n is the product of n copies of S^1. We call $S^1 \times D^2$ a *solid torus*.

Exercise 1.10 Show that $\partial(M \times N) = (\partial M \times N) \cup (M \times \partial N)$.

Remark 1.11 Clearly, $(\partial M \times N) \cap (M \times \partial N) = \partial M \times \partial N$. In particular, by applying the formula of Exercise 1.10 to $S^3 = \partial D^4 \approx \partial(D^2 \times D^2)$, we see that

$$S^3 = (S^1 \times D^2) \cup (D^2 \times S^1).$$

In other words, we can obtain the three-sphere by gluing two solid tori along their boundaries. The gluing map interchanges the meridian $\{1\} \times S^1$ and the longitude $S^1 \times \{1\}$ of $T^2 = \partial(S^1 \times D^2)$.

Another important operation on manifolds is the *connected sum*. Suppose that M and N are manifolds of the same dimension. Remove interiors of closed balls from M and N, and glue them along their sphere boundaries. The resulting manifold $M \# N$ only depends on which components of M and N we removed the balls from, up to homeomorphism. If $n > 2$, then $\pi_1(S^{n-1}) \cong 1$, and hence $\pi_1(M \# N)$ is the free product of $\pi_1(M)$ and $\pi_1(N)$ by the Seifert–van Kampen theorem.

A manifold of dimension two is also called a *surface*. For a positive integer g, let Σ_g be the connected sum of g copies of the two-torus T^2, and we write $\Sigma_0 := S^2$. The surface Σ_g is orientable; that is, two-sided. Here g is called the *genus* of the orientable surface Σ_g. We can obtain the non-orientable surface N_g by taking the connected sum of $g \geq 1$ copies of the real projective plane $\mathbb{RP}^2$. We now state the classification of compact surfaces without proof.

Theorem 1.12 *Every closed, connected surface is homeomorphic to either* Σ_g *for some* $g \geq 0$, *or* N_g *for some* $g \geq 1$. *Compact surfaces with boundary are obtained from these by removing finitely many open balls.*

The proof for triangulated surfaces can be found in most introductory textbooks; see, for example, Munkres [125]. The fact that every topological surface can be triangulated is a non-trivial result of Radó [150]; see also Moise [120] and Hatcher [59]. This is not true in higher dimensions, as there are topological manifolds in every dimension at least five that cannot be triangulated by the work of Manolescu [104]. Surprisingly, the proof of this result relies on a gauge-theoretic invariant of three-manifolds called monopole Floer homology. In Section 5, we will give an overview of a closely related theory called Heegaard Floer homology.

Three-manifolds are much more complicated. While a lot is known about them, we do not have a complete classification as in dimension two and below. The most important three-manifold invariant is the fundamental group. Perelman proved the famous Poincaré Conjecture only 100 years after it was first formulated. This states that the only closed simply connected three-manifold is S^3. In fact, he proved the much stronger Geometrisation Conjecture of Bill Thurston: every closed three-manifold can be cut along embedded two-spheres and two-tori such that each of the resulting pieces caries one of eight special geometric structures. These include spherical, Euclidean, and hyperbolic. The most difficult to understand are the hyperbolic pieces, which are determined by their fundamental groups by Mostow rigidity.

In dimensions four and higher, there is no hope for obtaining a complete classification. This is due to the following two results:

Theorem 1.13 *For every finitely presented group G and integer $n \geq 4$, there exists a closed n-manifold M with $\pi_1(M) \cong G$.*

Proof Let $\langle x_1, \ldots, x_k \mid r_1, \ldots, r_l \rangle$ be a presentation of G. Consider the n-manifold X obtained by taking the connected sum of k copies of $S^1 \times S^{n-1}$. Then $\pi_1(X)$ is the free product of k copies of

$$\pi_1(S^1 \times S^{n-1}) \cong \pi_1(S^1) \times \pi_1(S^{n-1}) \cong \mathbb{Z};$$

that is, $\pi_1(X) \cong \langle x_1, \ldots, x_k \rangle$.

We are now going to change X to introduce the relations $r_1, \ldots, r_l$ using an operation called surgery that we will discuss in more detail later. Let γ_i be a curve in X freely homotopic to the relation $r_i \in \pi_1(X)$ for $i \in \{1, \ldots, l\}$. We can assume that the curves $\gamma_1, \ldots, \gamma_l$ are embedded, pairwise disjoint, and have neighbourhoods $N(\gamma_i)$ homeomorphic to $S^1 \times D^{n-1}$ by representing each generator x_i using curves of the form $S^1 \times \{p\}$ for some $p \in S^{n-1}$.

An application of the Seifert–van Kampen theorem shows that

$$\pi_1(X \setminus N(\gamma_1 \cup \cdots \cup \gamma_l)) \cong \pi_1(X).$$

Indeed, $\pi_1(N(\gamma_i)) \cong \mathbb{Z}$ is generated by $[\gamma_i]$, which is homotopic to the generator of

$$\pi_1(\partial N(\gamma_i)) \cong \pi_1(S^1 \times S^{n-2}) \cong \mathbb{Z},$$

where we have used the fact that $\pi_1(S^{n-2}) \cong 1$ as $n \geq 4$. Hence, gluing $N(\gamma_i)$ to $X \setminus N(\gamma_1 \cup \cdots \cup \gamma_l)$ does not change the fundamental group, as it introduces a new generator y_i and a new relation $y_i = w_i$ for an element

$$w_i \in \pi_1(X \setminus N(\gamma_1 \cup \cdots \cup \gamma_l)).$$

Finally, we glue a copy of $D^2 \times S^{n-2}$ to $\partial N(\gamma_i) \approx S^1 \times S^{n-2}$ such that ∂D^2 is identified with $S^1 \times \{p\}$ for each $i \in \{1, \dots, l\}$ for some $p \in S^{n-2}$. Again, by the Seifert–van Kampen theorem, this kills the homotopy class of γ_i. □

Remark 1.14 In contrast, note that the fundamental group of every closed three-manifold has a presentation with the same number of generators as relations and is hence not arbitrary. This holds since every closed three-manifold admits a Heegaard decomposition; see Chapter 3.5 and Exercise 3.30.

The following result is due to Adian [3] and Rabin [149]:

Theorem 1.15 *There is no algorithm to decide whether two finitely presented groups are isomorphic.*

Hence, to classify manifolds in dimension at least four, one has to put restrictions on the fundamental group. A natural choice is the class of simply connected manifolds. Simply connected topological four-manifolds were classified by Freedman; see Chapter 6.2.

1.2 Smooth Manifolds

We say that a map from $\mathbb{R}^n$ to $\mathbb{R}^m$ is *smooth* or C^∞ if it is infinitely differentiable; that is, its coordinate functions have continuous partial derivatives of arbitrarily high order. Smooth manifolds are topological manifolds with some extra structure that allows one to define smooth maps between them. We can think of a smooth n-manifold as being glued together from open subsets of $\mathbb{R}^n$ such that the gluing maps are all smooth with a smooth inverse (i.e., diffeomorphisms). Alternatively, we can think of it as a topological manifold with a set of local charts that form the pages of an atlas, and when converting coordinates from one chart to the other, we apply a smooth map. For example, a usual atlas of the surface of the Earth, which is topologically S^2, consists of pages each showing a planar region.

Definition 1.16 Let M be a topological n-manifold. A *chart* on M is a homeomorphism ϕ from an open subset U of M to an open subset of $\mathbb{R}^n$. A *smooth atlas* of M is a set $\mathcal{A}$ of charts $\{(U_i, \phi_i) : i \in \mathcal{I}\}$ such that $M = \bigcup_{i \in \mathcal{I}} U_i$ and the *transition function*

$$\phi_j \circ \phi_i^{-1} : \phi_i(U_i \cap U_j) \to \phi_j(U_i \cap U_j)$$

is smooth for every i, $j \in \mathcal{I}$. We say that the smooth atlases $\mathcal{A}$ and $\mathcal{A}'$ are *equivalent* if $\mathcal{A} \cup \mathcal{A}'$ is also a smooth atlas. A *smooth structure* on the topological manifold M is an equivalence class of smooth atlases. A *smooth* (or *differentiable*) *manifold* is a topological manifold endowed with a smooth structure.

Remark 1.17 In the preceding definition, we could replace the word 'smooth' with 'C^k-differentiable' to define *C^k-differentiable manifolds*. However, there is a bijection between C^k-differentiable and smooth structures on any topological manifold for $k \geq 1$, so we will only study smooth manifolds. We will use the terms 'local coordinates' and 'chart' interchangeably.

If we require the transition functions to be analytic, we obtain the class of *real analytic manifolds*. If we are given charts to $\mathbb{C}^k$ with biholomorphic transition functions, we obtain *complex manifolds*. A one-dimensional complex manifold is called a *Riemann surface*.

If the transition maps are piecewise linear, we obtain the notion of *piecewise linear* or *PL manifolds*. A triangulation of a topological manifold is a homeomorphism with the polyhedron of a simplicial complex. Recall that the link of a vertex v of a triangulation is the boundary of the union of the closed simplices incident to v. A PL structure corresponds to a triangulation where the link of every vertex is a sphere.

Exercise 1.18 Show that stereographic projections from the North and South poles form charts of an atlas on S^n. Its equivalence class is the *standard smooth structure on S^n*.

Every smooth manifold can be triangulated; see Munkres [124]. However, not every triangulable manifold admits a smooth structure, even if we assume that the boundary of the link of each vertex is a sphere.

Definition 1.19 Let M and N be smooth manifolds with smooth atlases $\{(U_i, \phi_i) : i \in \mathcal{I}\}$ and $\{(V_j, \psi_j) : j \in \mathcal{J}\}$. We say that the map $f : M \to N$ is C^r for $r \in \mathbb{N} \cup \{\infty\}$ if the map

$$\psi_j \circ f \circ \phi_i^{-1} : \phi_i(U_i \cap f^{-1}(V_j)) \to \psi_j(f(U_i) \cap V_j)$$

is r-times continuously differentiable for every $i \in \mathcal{I}$ and $j \in \mathcal{J}$. (This does not depend on the choice of atlases.) We denote the set of C^r maps from M to N by $C^r(M,N)$. We will often call C^∞ maps *smooth*.

We can endow $C^r(M,N)$ with several topologies, of which we will now define the weak C^r topology:

Definition 1.20 Let M and N be smooth manifolds, and let $f \in C^r(M,N)$ for some $r \in \mathbb{N}$. Furthermore, let (U,ϕ) be a chart on M and (V,ψ) be a chart on N, and $K \subset U$ a compact set such that $f(K) \subset V$. For $\varepsilon > 0$, the set of $g \in C^r(M,N)$ for which

$$\left|\mathrm{D}^k(\psi \circ g \circ \phi^{-1})(x) - \mathrm{D}^k(\psi \circ f \circ \phi^{-1})(x)\right| < \varepsilon$$

for every $x \in \phi(K)$ and $k \leq r$ form a sub-basis of neighbourhoods of f in the weak C^r topology. The *weak C^∞ topology* on $C^\infty(M,N)$ is the union of the subspace topologies induced from $C^r(M,N)$ for $r \in \mathbb{N}$.

The coordinate-independent notion of Taylor polynomial of a map between manifolds is called a *jet*.

Definition 1.21 Let M and N be smooth manifolds and $r \in \mathbb{N}$. Fix a point $p \in M$ and an integer $k \in \{0,\ldots,r\}$. We say that the maps $g \in C^r(U,N)$ and $h \in C^r(V,N)$ for open neighbourhoods U and V of p are *equivalent* if $g(p) = h(p)$ and all their partial derivatives up to order k agree at p in charts about p and $g(p)$. Then the k-jet of $f \in C^r(M,N)$ at p, which we denote by $j^k f(p)$, is the equivalence class of f.

For $k \leq r$, the k-jets of maps in $C^r(M,N)$ form a bundle over M whose total space we denote by $J^k(M,N)$. See Appendix A for an introduction to fibre bundles. If we take the inverse limit of the spaces $J^k(M,N)$ under the forgetful maps $J^{k+1}(M,N) \to J^k(M,N)$, we obtain $J^\infty(M,N)$.

Notice that $j^k f$ for $f \in C^r(M,N)$ is a section of $J^k(M,N)$. We endow $\Gamma(J^k(M,N))$, the space of sections of the bundle $J^k(M,N)$, with the compact-open topology, whose sub-basis consists of sections that map a given compact set in M into a given open subset of $J^k(M,N)$. Then the *weak C^k topology* on $C^r(M,N)$ is obtained by pulling back the compact-open topology on $\Gamma(J^k(M,N))$ under the map $f \mapsto j^k f$.

Smooth manifolds with smooth maps between them form a category. An equivalence in this category is called a diffeomorphism.

Definition 1.22 The map $\phi\colon M \to N$ is a *diffeomorphism* if it is a homeomorphism such that both ϕ and ϕ^{-1} are smooth.

Two diffeomorphisms ϕ_0, $\phi_1\colon M \to N$ are *pseudoisotopic* if there is a diffeomorphism

$$\Phi\colon M \times I \to N \times I$$

such that $\Phi(x,0) = (\phi_0(x),0)$ and $\Phi(x,1) = (\phi_1(x),1)$ for every $x \in M$. We say that Φ is an *isotopy* if it is also *level-preserving*; that is, $\Phi(x,t) \in N \times \{t\}$ for every $x \in M$. Then ϕ_0 and ϕ_1 are called *isotopic*.

Given an isotopy Φ, we will use the notation $\Phi_t(x) := \Phi(x,t)$ for $t \in I$. Then we can think of Φ as a smooth one-parameter family of diffeomorphisms Φ_t connecting ϕ_0 and ϕ_1. In fact, if we endow the space of diffeomorphisms $\mathrm{Diff}(M,N)$ from M to N with the C^∞ topology, then two diffeomorphisms are isotopic if and only if they lie in the same path component of $\mathrm{Diff}(M,N)$. The proof of this involves deforming a continuous path of diffeomorphisms to a smooth one, which we omit.

We now define tangent vectors and tangent and cotangent spaces.

Definition 1.23 Let M be a smooth n-manifold and $p \in M$ a point. Given smooth curves $\gamma_1, \gamma_2\colon (-\varepsilon, \varepsilon) \to M$ such that $\gamma_1(0) = \gamma_2(0) = p$, we say that γ_1 and γ_2 are *equivalent* if their velocity vectors $(\phi \circ \gamma_1)'(0)$ and $(\phi \circ \gamma_2)'(0)$ are the same in any chart (and hence all charts) (U,ϕ) with $p \in U$. A *tangent vector* of M at p is an equivalence class of such curves.

We denote by T_pM the set of all tangent vectors of M at p, and call it the *tangent space* of M at p. This is a vector space, where the operations are the natural ones given in a coordinate chart; for example, $[\gamma_1] + [\gamma_2] = [\gamma_3]$ if

$$(\phi \circ \gamma_1)'(0) + (\phi \circ \gamma_2)'(0) = (\phi \circ \gamma_3)'(0)$$

for some chart (U,ϕ) about p. Finally, we write

$$TM := \bigcup_{p \in M} T_pM$$

for the *tangent bundle* of M. This is a vector bundle with base space M. A *vector field* on M is a section of the tangent bundle TM.

The *cotangent bundle* T^*M of M is obtained by taking the union of the dual spaces $T_p^*M := (T_pM)^*$ for $p \in M$.

To define the topology on TM, we choose an atlas $\mathcal{A} = \{(U_i, \phi_i)\colon i \in \mathcal{I}\}$ on M and glue together the product bundles $U_i \times \mathbb{R}^n$ for $i \in \mathcal{I}$, as follows. For i,

$j \in \mathcal{I}$, a point $x \in U_i \cap U_j$, and a vector $v \in \mathbb{R}^n$, we identify $(x, v) \in U_i \times \mathbb{R}^n$ with

$$\left(x, D(\phi_j \circ \phi_i^{-1})(\phi_i(x))(v)\right) \in U_j \times \mathbb{R}^n,$$

where $D(\phi_j \circ \phi_i^{-1})(\phi_i(x))$ is the Jacobian of $\phi_j \circ \phi_i^{-1}$ at $\phi_i(x)$.

Using the language of jets, a tangent vector at p is a one-jet $j^1\gamma(0)$ for $\gamma \in C^1(\mathbb{R}, M)$ such that $\gamma(0) = p$.

Exercise 1.24 Show that $J^1(\mathbb{R}, M) = \mathbb{R} \times TM$ and $J^1(M, \mathbb{R}) = T^*M \times \mathbb{R}$ as bundles over M.

A subset of a smooth n-manifold is called a submanifold if it is locally modelled on $(\mathbb{R}^n, \mathbb{R}^k)$ for some $k \leq n$:

Definition 1.25 Let M be a smooth n-manifold and $k \leq n$. We say that $N \subset M$ is a smooth k-dimensional *submanifold* if there is a chart (U, ϕ) about each $p \in N$ such that $\phi(N \cap U) = \phi(U) \cap \mathbb{R}^k$. The charts $(N \cap U, \phi|_{N \cap U})$ define an atlas of N, which becomes a smooth k-manifold.

If N is a smooth submanifold of M, then TN is a sub-bundle of the restriction $TM|_N$. The *normal bundle* of N in M is defined as the quotient

$$\nu_{N \subset M} := (TM|_N)/TN.$$

Definition 1.26 Let S be a smooth submanifold of M. We write E for the total space of the normal bundle $\nu_{S \subset M}$ and $0_S \colon S \longrightarrow E$ for the zero section. Then we say that the open subset $N(S) \subset M$ is a *tubular* or *regular neighbourhood* of S if there is a diffeomorphism $\varphi \colon E \longrightarrow N(S)$ such that $\varphi \circ 0_S = \mathrm{Id}_S$.

Proposition 1.27 *Every smooth submanifold has a tubular neighbourhood.*

For a proof, see Hirsch [65].

Definition 1.28 Let N and N' be smooth submanifolds of the manifold M. We say that N and N' are *transverse* (or *intersect transversely*) if, at each intersection point $p \in N \cap N'$, we have

$$T_pN + T_pN' = T_pM.$$

If N and N' are transverse, then $N \cap N'$ is a smooth manifold. Furthermore,

$$\dim(N \cap N') = \dim(N) + \dim(N') - \dim(M),$$

which follows from the observation that $T_p(N \cap N') = T_pN \cap T_pN'$ for $p \in N \cap N'$, and the dimension formula for the intersection of two linear subspaces of a vector space.

Definition 1.29 If N and N' are smooth submanifolds of M, then they are said to be *ambient isotopic* (or simply *isotopic*) if there is an isotopy $\Phi\colon M \times I \to M \times I$ such that $\Phi_0 = \mathrm{Id}_M$ and $\Phi_1(N) = N'$.

Definition 1.30 Let $f\colon M \to N$ be a smooth map. The *differential* of f at $p \in M$ is a linear map

$$\mathrm{d}f_p\colon T_pM \to T_{f(p)}N,$$

defined as follows. Let γ be a curve representing a vector $v \in T_pM$. Then $\mathrm{d}f_p(v)$ is the tangent vector of N at $f(p)$ represented by $f \circ \gamma$. The maps $\mathrm{d}f_p$ for $p \in M$ assemble to a morphism of vector bundles $\mathrm{d}f\colon TM \to TN$.

Let $C^\infty(M)$ denote the vector space of smooth functions $f\colon M \to \mathbb{R}$. Given a tangent vector $v \in T_pM$ represented by a curve $\gamma\colon (-\varepsilon,\varepsilon) \to M$ and a smooth function $f \in C^\infty(M)$, we can define the *directional derivative* vf as $(f \circ \gamma)'(0)$. This is independent of the choice of representative γ. The differential $\mathrm{d}f$ of the function f is a section of the cotangent bundle T^*M. By definition, $\mathrm{d}f(v) = vf$.

From the cotangent bundle T^*M, we can form the bundle $\Lambda^k T^*M$ for every $k \in \mathbb{N}$, whose fibre over $p \in M$ is the kth exterior power of T_p^*M. Its sections are called *k-forms* and are denoted by $\Omega^k(M)$. The differential of functions extends to a linear map

$$\mathrm{d}^k\colon \Omega^k(M) \to \Omega^{k+1}(M),$$

as follows. Consider the multi-index $I := (i_1,\dots,i_k)$, where $1 \le i_1 < i_2 < \cdots < i_k \le n$ are integers. In local coordinates $(x_1,\dots,x_n)$, consider the k-form $\mathrm{d}x^I := \mathrm{d}x_{i_1} \wedge \cdots \wedge \mathrm{d}x_{i_k}$. For a smooth function f, we let

$$\mathrm{d}^k\left(f\mathrm{d}x^I\right) := \sum_{i=1}^{n} \frac{\partial f}{\partial x_i}\mathrm{d}x_i \wedge \mathrm{d}x^I,$$

and extend d^k to k-forms linearly. Since $\mathrm{d}^{k+1} \circ \mathrm{d}^k = 0$, the graded vector space

$$\Omega^*(M) := \bigoplus_{k=0}^{\infty} \Omega^k(M)$$

with the endomorphism $\mathrm{d} := \sum_{k=0}^{\infty} \mathrm{d}^k$ is a cochain complex, whose cohomology is called the *de Rham cohomology* $H^*_{dR}(M)$ of M. Note that $\Omega^k(M) = 0$ for

$k > n$, since $\Lambda^k V = \{0\}$ for an n-dimensional vector space V and $k > n$. The main result of de Rham theory is the following.

Theorem 1.31 *If M is a closed, smooth n-manifold, then*

$$H^*_{dR}(M) \cong H^*(M;\mathbb{R}).$$

For a proof, see Bott and Tu [11].

Proposition 1.32 *A smooth n-manifold M is orientable if and only if the line bundle $\Lambda^n T^* M$ is trivial. An orientation of M corresponds to a nowhere vanishing section of $\Lambda^n T^* M$, called a* volume form.

See Bott and Tu [11, proposition 3.2] for a proof. If $\omega \in \Omega^n(M)$ is a volume form and $p \in M$, then we say that a basis $v_1, \dots, v_n$ of $T_p M$ is *positive* if $\omega_p(v_1, \dots, v_n) > 0$, and is *negative* otherwise. This is consistent with our intuitive notion of orientability. Alternatively, a smooth n-manifold is orientable if and only if it admits an atlas where all the transition maps are orientation-preserving, in the sense that they have differentials in $\mathrm{GL}_+(n,\mathbb{R})$.

Let us write $C^\infty_p(M)$ for the vector space of germs of smooth functions on M at p. This consists of equivalence classes of smooth functions on M defined in a neighbourhood of p such that $f \sim g$ if there exists an open set U in M containing p such that $f|_U = g|_U$. A *derivation* D at $p \in M$ is a linear functional on $C^\infty_p(M)$ that satisfies the Leibniz rule; that is,

$$\mathrm{D}(fg) = \mathrm{D}(f)g + f\mathrm{D}(g).$$

Then there is a bijection between $T_p M$ and the space of derivations at p. Hence, alternatively, we could have defined tangent vectors as derivations. Vector fields correspond to linear transformations on $C^\infty(M)$ that satisfy the Leibniz rule. For example, given a coordinate chart $\phi \colon U \to \mathbb{R}^n$ on M, we obtain the *coordinate vector fields* $\partial/\partial x_i$ on U for $i \in \{1, \dots, n\}$ by letting

$$(\partial/\partial x_i)(f) := \partial(f \circ \phi^{-1})/\partial x_i$$

for a function $f \in C^\infty(U,\mathbb{R})$.

Differential topology is the study of the category of smooth manifolds and smooth maps between them. One of the main questions is the classification of smooth structures on topological manifolds.

Theorem 1.33 *Every topological manifold of dimension at most three has a unique smooth structure, up to diffeomorphism.*

Moise [119][120] showed that every two-manifold and three-manifold admits a unique PL structure; see also Hamilton [56]. For the uniqueness of smoothing of the PL structure, see Hirsch and Mazur [66] and Thurston [171].

Things suddenly change in dimension four. There are topological manifolds that admit no smooth structure, and there are some that admit infinitely many. This will be the subject of Chapter 6.

Surprisingly, things get somewhat easier in dimensions five and higher, at least once one restricts the fundamental group and considers only, say, simply connected manifolds. This is because of the h-cobordism theorem that we will discuss in Section 2.3. At the heart of the proof of the h-cobordism theorem lies the *Whitney trick*. This is a geometric operation aimed at cancelling pairs of intersection points of opposite sign between two transverse submanifolds of complementary dimensions, and hence it can be used to realise algebraic intersection numbers as geometric intersection numbers. Using this, Kervaire and Milnor [82] determined the set of smooth structures on all spheres of dimension at least five in terms of stable homotopy groups of spheres. For example, there is a unique smooth structure on S^5 and S^6, but there are 28 on S^7! We will prove the latter in Section 2.8.

Given a smooth manifold M, a non-diffeomorphic smooth structure on M is called *exotic*. It is not known whether there is an exotic smooth structure on S^4. This is called the *smooth four-dimensional Poincaré conjecture.*

1.3 Embeddings, Immersions, and Submersions

Manifolds were first considered as subsets of some Euclidean space, and defined as common zero sets of a collection of smooth functions. In this section, we will show that every abstract smooth manifold defined using atlases embeds into some Euclidean space.

Definition 1.34 A smooth map $f\colon M \to N$ is an *immersion* if its differential $\mathrm{d}f_p\colon T_pM \to T_{f(p)}N$ is injective for every $p \in M$.

We say that f is an *embedding* if it is an injective immersion that is a homeomorphism onto its image.

We will use the notation $f\colon M \looparrowright N$ for immersions. An immersion f is locally an embedding, and its image is locally a submanifold of N. However, it might have self-intersections; that is, distinct points $p, q \in M$ such that $f(p) = f(q)$. For example, consider a map from S^1 to $\mathbb{R}^2$ parametrising a figure eight curve in the plane.

Not every injective immersion is an embedding. The map $f\colon \mathbb{R} \to \mathbb{R}^2/\mathbb{Z}^2$ given by $f(t) = tv$ for $v = (1,s)$ and $s \in \mathbb{R} \setminus \mathbb{Q}$ has dense image in the torus $T^2 = \mathbb{R}^2/\mathbb{Z}^2$. So, f is not a homeomorphism onto its image when that is endowed with the subspace topology. However, every injective immersion of a compact manifold is an embedding. If f is an embedding, then $f(M)$ is a submanifold of N.

Definition 1.35 We say that the embeddings f_0, $f_1\colon M \to N$ are *isotopic* if there is a smooth map $F\colon M \times I \to N$ such that $F_t := F(-,t)$ is an embedding for every $t \in I$, and $F_i = f_i$ for $i \in \{0,1\}$. We call F an *isotopy.*

We can think of an isotopy as a path in the space $\mathrm{Emb}(M,N)$ of embeddings of M into N, considered as a subspace of $C^\infty(M,N)$. In Definition 1.29, we introduced the notion of ambient isotopy for two submanifolds. There is an analogous notion for embeddings. According to the isotopy extension theorem, every isotopy can be extended to an ambient isotopy.

Theorem 1.36 (Isotopy extension theorem) *Let M be a compact submanifold of N. Suppose that $F\colon M \times I \to N$ is an isotopy such that F_0 is the embedding of M into N, where $F_t := F(-,t)$ for $t \in I$. Then there is a smooth map $G\colon N \times I \to N$ such that $G_t := G(-,t)$ is a diffeomorphism, $G_0 = \mathrm{Id}_N$, and*

$$G_t \circ F_0 = F_t$$

for every $t \in I$.

Proof We obtain G by integrating a vector field v on $N \times I$. This will be of the form $v_N + \partial/\partial t$, where v_N is tangent to the N-direction. We first define v_N along the submanifold

$$L := \{(F_t(x),t)\colon (x,t) \in M \times I\}$$

of $N \times I$. For $(x,t) \in M \times I$, we let $v_N(F_t(x),t)$ be the velocity vector of the curve $\gamma_x(t) := F_t(x)$ at t. Let $N(L)$ be a tubular neighbourhood of L in $N \times I$. We then extend v_N to $N(L)$ and multiply it with a fibrewise bump function on $N(L)$ that is 1 along L and 0 on a neighbourhood of $(N \times I) \setminus N(L)$. We finally let $v_N = 0$ on $(N \times I) \setminus N(L)$.

By construction, v is tangent to the curves $t \mapsto (\gamma_x(t),t)$ for $(x,t) \in M \times I$. Given $(x,t) \in N \times I$, we obtain $(G(x,t),t)$ by following the flow of v starting from $(x,0)$ until we reach $N \times \{t\}$. □

Remark 1.37 As the space of vector fields v tangent to $F(\{x\} \times I)$ for every $x \in M$ and of the form $v_N + \partial/\partial t$ in the proof of Theorem 1.36 is convex and hence contractible, the diffeomorphism $(G(x,t),t)$ of $N \times I$ is unique up to isotopy.

Corollary 1.38 (Homogeneity lemma) *Let N be a connected manifold and x, $y \in N$. Then there is an isotopy $G\colon N \times I \to N$ such that $G_0 = Id_N$ and $G_1(x) = y$.*

Proposition 1.39 *For every smooth n-manifold M, there is an embedding*

$$f\colon M \hookrightarrow \mathbb{R}^a$$

for some $a \in \mathbb{N}$.

Proof We only give the proof when M is compact. Then there are charts $\varphi_i\colon U_i \to \mathbb{R}^n$ for $i \in \{1,\dots,k\}$ such that each U_i is diffeomorphic to the ball B^n, and there are concentric balls $U_i'' \subsetneq U_i' \subsetneq U_i$ with $M = \cup_{i=1}^k U_i''$. There are smooth functions $\mu_i\colon M \to I$ and $\lambda_i\colon M \to I$ such that

- $\mu_i|_{U_i'} \equiv 1$ and $\mu_i|_{M \setminus U_i} \equiv 0$, and
- $\lambda_i|_{U_i''} \equiv 1$ and $\lambda_i|_{M \setminus U_i'} \equiv 0$.

Then the maps

$$\psi_i := \mu_i \varphi_i\colon M \to \mathbb{R}^n$$

are smooth for $i \in \{1,\dots,k\}$. Furthermore, if we set $\psi = (\psi_1,\dots,\psi_k)$, then $\psi\colon M \to \mathbb{R}^{nk}$ is an immersion. Indeed, for $p \in M$, there is an $i \in \{1,\dots,k\}$ such that $p \in U_i''$, and $(\mathrm{d}\psi_i)_p$ has rank n.

To lift ψ to an embedding, let $\Lambda := (\lambda_1,\dots,\lambda_k)$, and set

$$f := (\psi,\Lambda)\colon M \to \mathbb{R}^{nk} \times \mathbb{R}^k.$$

Suppose that $f(x) = f(y)$ for $x \neq y \in M$. Then $x \in U_i''$ for some $i \in \{1,\dots,k\}$. As $\psi_i(x) = \psi_i(y)$, and since $\psi_i = \mu_i\varphi_i = \varphi_i$ is injective in U_i', we have $y \notin U_i'$. But then $\lambda_i(y) = 0$ and $\lambda_i(x) = 1$, which is a contradiction. □

Definition 1.40 Let $f\colon M \to N$ be a smooth map. We say that $p \in M$ is a *regular point* of f if $\mathrm{rk}(\mathrm{d}f_p) = \dim(N)$, and is a *critical point* otherwise. The point $q \in N$ is a *regular value* of f if every point of $f^{-1}(q)$ is regular, and is a *critical value* otherwise. The map f is a *submersion* if every $p \in M$ is regular.

If q is a regular value, then $f^{-1}(q)$ is a submanifold of M by the implicit function theorem.

Exercise 1.41 (Ehresmann's fibration lemma) Let $f\colon M \to N$ be a submersion. Show that if f is proper; that is, $f^{-1}(K)$ is compact for every $K \subseteq N$ compact, then M is a fibre bundle over N with fibre $f^{-1}(\{q\})$ for $q \in N$. Give a counterexample when f is not proper.

A subset of a smooth manifold is said to be *measure zero* if its intersection with any coordinate chart has measure zero. This is well defined since diffeomorphisms map measure zero sets to measure zero sets, and change of coordinate maps are diffeomorphisms. The following result of Sard plays an important role in differential topology.

Theorem 1.42 (Sard's theorem) *Let $f\colon M \to N$ be a smooth map. Then the set of critical values of f forms a measure zero subset of N.*

For a proof, see Milnor [116]. As a special case, when $\dim(M) < \dim(N)$, we obtain that $\mathrm{Im}(f)$ has measure zero in N.

It is an important question in differential topology what the smallest a is such that we can embed or immerse a given smooth n-manifold into $\mathbb{R}^a$.

Theorem 1.43 *Every smooth n-manifold M can be embedded into $\mathbb{R}^{2n+1}$ and immersed into $\mathbb{R}^{2n}$.*

Proof Again, we only prove the case when M is compact. By Proposition 1.39, there is an embedding $f\colon M \hookrightarrow \mathbb{R}^a$ for some $a \in \mathbb{N}$. For $v \in S^{a-1}$, let p_v be orthogonal projection onto $v^\perp$. Our goal is to find a direction $v \in S^{a-1}$ such that $p_v \circ f$ is also an embedding or immersion.

For $p_v \circ f$ to be an immersion, $\mathrm{d}(p_v \circ f)$ has to have rank n everywhere. To ensure this, v should not be tangent to $\mathrm{Im}(f)$. This means that

$$v \neq \frac{\mathrm{d}f(w)}{|\mathrm{d}f(w)|}$$

for w a non-zero tangent vector of M. The *unit tangent bundle*

$$STM := (TM \setminus 0_M)/\mathbb{R}^*$$

of M is $(2n-1)$-dimensional, as it has fibre S^{n-1}, where 0_M is the zero section. Hence, by applying Sard's theorem to $\mathrm{d}f/|\mathrm{d}f|\colon STM \to S^{a-1}$, if $a > 2n$, for

v outside a measure zero subset of S^{a-1}, the map $p_v \circ f$ is an immersion. Repeatedly projecting in this way, we obtain an immersion of M into $\mathbb{R}^{2n}$.

To make sure $p_v \circ f$ is also injective, consider the map

$$\varphi\colon (M \times M) \setminus \Delta \to S^{a-1}$$
$$(x, y) \mapsto \frac{f(x) - f(y)}{|f(x) - f(y)|}$$

for $x \neq y \in M$, where

$$\Delta = \{(x, x)\colon x \in M\}$$

is the diagonal. By Sard's theorem, the image of φ has measure zero in S^{a-1} whenever $a - 1 > \dim(M \times M) = 2n$. If $v \in S^{a-1} \setminus \operatorname{Im}(\varphi)$, then $p_v \circ f$ is injective. Furthermore, if v also avoids the image of STM, the map $p_v \circ f$ is also an immersion, and is hence an embedding as M is compact. Repeating this process, we obtain an embedding of M into $\mathbb{R}^{2n+1}$. □

The preceding proof can be modified to show that, in fact, embeddings of an n-manifold into $\mathbb{R}^{2n+1}$ are dense in the space of all smooth maps, which we endow with the C^∞ topology. Similarly, the set of immersions into $\mathbb{R}^{2n}$ is also dense. In other words, any smooth map of an n-manifold into $\mathbb{R}^{2n+1}$ can be perturbed into an embedding, and every smooth map into $\mathbb{R}^{2n}$ can be perturbed into an immersion.

We will see in Section 2.2 that Theorem 1.43 can be improved, and every n-manifold embeds into $\mathbb{R}^{2n}$ for $n > 0$ and immerses into $\mathbb{R}^{2n-1}$ for $n > 1$. However, the subsets of embeddings and immersions are no longer dense in these dimensions.

Given a manifold, it is a hard problem to determine the smallest a such that it embeds or immerses into $\mathbb{R}^a$. The *immersion conjecture* states that if $\alpha(n)$ is the number of ones in the binary expansion of n, then every n-manifold immerses into $\mathbb{R}^{2n-\alpha(n)}$. This has been shown by Cohen [25].

1.4 Connected Sums

We have already encountered the connected sum operation for topological manifolds. The corresponding operation in the smooth category is more subtle.

Definition 1.44 We define the *connected sum* of two smooth, oriented, connected n-manifolds M_1 and M_2 as follows. Choose embeddings $e_i\colon D^n \hookrightarrow M_i$

for $i \in \{1,2\}$ such that e_1 is orientation-preserving and e_2 is orientation-reversing. We obtain $M_1 \# M_2$ from the disjoint union

$$(M_1 \setminus \{e_1(0)\}) \sqcup (M_2 \setminus \{e_2(0)\})$$

by identifying $e_1(tv)$ with $e_2((1-t)v)$ for each unit vector $v \in S^{n-1}$ and $t \in (0,1)$. We orient $M_1 \# M_2$ compatibly with M_1 and M_2.

This is independent of the choice of embeddings e_1 and e_2 up to diffeomorphism by the work of Cerf [19] and Palais [143], who showed that, for any two embeddings $e, e' \colon D^n \to M$, there is a diffeomorphism $f \colon M \to M$ such that $e' = f \circ e$. Furthermore, the connected sum operation is commutative and associative up to diffeomorphism, and S^n is an identity element.

Example 1.45 The connected sum operation does not satisfy the cancellation law: If $\overline{\mathbb{CP}}^2$ denotes $\mathbb{CP}^2$ with the reverse of the complex orientation, then one can show using Kirby calculus (Exercise 6.3) that

$$\mathbb{CP}^2 \# \mathbb{CP}^2 \# \overline{\mathbb{CP}}^2 \approx \mathbb{CP}^2 \# (S^2 \times S^2),$$

where '$\approx$' denotes diffeomorphism. However,

$$\mathbb{CP}^2 \# \overline{\mathbb{CP}}^2 \not\approx S^2 \times S^2,$$

as they have non-isomorphic cohomology rings.

There are 28 smooth manifolds homeomorphic to S^7 up to diffeomorphism, and they form a group under connected sum. Hence, there are smooth seven-manifolds M and N such that $M \# N \approx S^n$, but $M \not\approx S^n$. In contrast, we have the following result of Mazur, whose proof is particularly ingenious.

Theorem 1.46 *If $M \# N \approx S^n$, then M is* homeomorphic *to S^n.*

Proof Consider the infinite connected sum

$$X := M \# N \# M \# N \# \cdots .$$

Since $M \# N \approx S^n$, we have

$$X \approx (M \# N) \# (M \# N) \# \cdots \approx S^n \# S^n \# \cdots \approx \mathbb{R}^n.$$

Here, the last diffeomorphism corresponds to the decomposition

$$\mathbb{R}^n = \bigcup_{k=0}^{\infty} \{x \in \mathbb{R}^n : |x| \in [k, k+1]\}.$$

On the other hand, since $N \# M \approx S^n$, we have

$$X \approx M \# (N \# M) \# (N \# M) \# \cdots \approx M \# S^n \# S^n \# \cdots \approx M \# \mathbb{R}^n.$$

So $\mathbb{R}^n \approx M \# \mathbb{R}^n$. Since $M \approx M \# S^n$, if $p \in M$ is an arbitrary point, then $M \setminus \{p\} \approx M \# \mathbb{R}^n \approx \mathbb{R}^n$. Hence M is homeomorphic to S^n. □

Boundary connected sum is a closely related operation for manifolds with boundary.

Definition 1.47 Let W_1 and W_2 be oriented $(n+1)$-manifolds with connected boundary. Let $H^{n+1} := D^{n+1} \cap \mathbb{R}^{n+1}_+$ be a half-disk. Choose embeddings $e_i\colon (H^{n+1}, D^n) \hookrightarrow (W_i, \partial W_i)$ for $i \in \{1,2\}$ such that $e_2 \circ e_1^{-1}$ is orientation-reversing. We obtain the *boundary connected sum* $W_1 \natural W_2$ from

$$(W_1 \setminus \{e_1(0)\}) \sqcup (W_2 \setminus \{e_2(0)\})$$

by identifying $e_1(tv)$ with $e_2((1-t)v)$ for every $v \in S^n \cap \mathbb{R}^{n+1}_+$ and $t \in (0,1)$.

Note that

$$\partial(W_1 \natural W_2) = W_1 \# W_2.$$

1.5 Morse Theory

We can learn a lot about smooth manifolds by studying critical points of certain nice generic functions on them, which are called Morse. We write $C^\infty(M) := C^\infty(M,\mathbb{R})$, and let $f \in C^\infty(M)$. By Definition 1.40, we say that $p \in M$ is a *critical point* of f if $\mathrm{d}f_p = 0$; that is, if $vf = 0$ for every $v \in T_pM$. Furthermore, $c \in \mathbb{R}$ is a *critical value* of f if there is a critical point p of f with $f(p) = c$, and is a *regular value* otherwise.

Definition 1.48 Let $f \in C^\infty(M)$ be a smooth function on the n-manifold M. We say that the critical point p is *non-degenerate* if, in a local coordinate system $(x_1, \dots, x_n)$ about p, the *Hessian*

$$\left(\frac{\partial^2 f}{\partial x_i \partial x_j}(\underline{0}) \right)_{i,j=1,\dots,n}$$

is non-degenerate. The function f is called *Morse* if all of its critical points are non-degenerate.

We will denote the set of critical points of a smooth function f by $\mathrm{Crit}(f)$. The non-degeneracy condition is equivalent to requiring that the section $df\colon M \to T^*M$ is transverse to the 0-section at p. There exists a local coordinate system about each non-degenerate critical point in which the function has a particularly nice form. More specifically, if $(x_1,\ldots,x_n)$ are local coordinates, then $\underline{0}$ is a critical point of f if and only if $\frac{\partial f}{\partial x_i}(\underline{0}) = 0$ for $i \in \{1,\ldots,n\}$. Non-degeneracy of the critical point amounts to non-degeneracy of the Hessian, which is a real symmetric bilinear form, and hence is diagonalisable as $\sum_{i=1}^n \varepsilon_i x_i^2$ for $\varepsilon_i \in \{1,-1\}$ by Sylvester's law of inertia. The Morse lemma states that there is a local coordinate system about a non-degenerate critical point in which the Hessian is diagonal with entries ± 1, and all higher-order terms of the Taylor series vanish.

Lemma 1.49 (Morse lemma) *Let p be a non-degenerate critical point of the smooth function $f \in C^\infty(M)$. Then there are local coordinates $\mathbf{x} = (x_1,\ldots,x_n)$ about p such that*

$$f(\mathbf{x}) = f(\mathbf{0}) - x_1^2 - \cdots - x_i^2 + x_{i+1}^2 + \cdots + x_n^2$$

for some $i \in \{0,\ldots,n\}$.

We call i the *index* of the critical point p and denote the set of index i critical points of f by $\mathrm{Crit}_i(f)$.

Proof We follow the exposition of Milnor [114]. By choosing a chart about p, we can assume that f is defined in a convex neighbourhood U of $\mathbf{0} \in \mathbb{R}^n$. Furthermore, by replacing f with $f - f(\mathbf{0})$, we can assume that $f(\mathbf{0}) = 0$. Then

$$f(\mathbf{x}) = \int_0^1 \frac{\mathrm{d}f}{\mathrm{d}t}(t\mathbf{x})\mathrm{d}t = \int_0^1 \sum_{i=1}^n \frac{\partial f}{\partial x_i}(t\mathbf{x})x_i \mathrm{d}t = \sum_{i=1}^n x_i \int_0^1 \frac{\partial f}{\partial x_i}(t\mathbf{x})\mathrm{d}t,$$

where we used the fundamental theorem of calculus and $f(\mathbf{0}) = 0$ in the first step. Repeating the same computation with $g_i(\mathbf{x}) := \int_0^1 \frac{\partial f}{\partial x_i}(t\mathbf{x})\mathrm{d}t$ in place of $f(\mathbf{x})$ and noting that $g_i(\mathbf{0}) = \frac{\partial f}{\partial x_i}(\mathbf{0}) = 0$, we obtain functions $H_{i,j}$ such that

$$f(\mathbf{x}) = \sum_{i,j=1}^n x_i x_j H_{i,j}(\mathbf{x}). \tag{1.1}$$

We can further assume that $H_{i,j} = H_{j,i}$ by replacing $H_{i,j}$ with $\frac{1}{2}(H_{i,j} + H_{j,i})$. Then $H_{i,j}(\mathbf{0}) = \frac{1}{2}\frac{\partial^2 f}{\partial x_i \partial x_j}(\mathbf{0})$.

We now adapt the proof of Sylvester's law of inertia to find the desired coordinate system. Suppose that there is a coordinate system $\mathbf{u} = (u_1,\ldots,u_n)$ about $\mathbf{0}$ in which

$$f(\mathbf{u}) = \sum_{i<r} \pm u_i^2 + \sum_{i,j=r}^{n} u_i u_j H_{i,j}(\mathbf{u}),$$

and $H_{i,j} = H_{j,i}$. This is true for $r = 1$ by Equation (1.1), and the main result follows from this by induction on r, as follows.

By non-degeneracy of the Hessian of f, there are $i, j \geq r$ such that $H_{i,j}(\mathbf{0}) \neq 0$. If $i \neq j$, then let $y_i = u_i + u_j$, $y_j = u_i - u_j$, and $y_k = u_k$ for $k \notin \{i, j\}$, giving

$$2u_i u_j H_{i,j} = \frac{1}{2}(y_i^2 - y_j^2)H_{i,j}.$$

Hence, we can assume that $H_{r,r}(\mathbf{0}) \neq 0$ by possibly reindexing the coordinates.

Write $g(\mathbf{u}) := \sqrt{|H_{r,r}(\mathbf{u})|}$. This is a smooth, positive function in some neighbourhood U' of $\mathbf{0}$ in U. We now set

$$v_r(\mathbf{u}) := g(\mathbf{u})\left(u_r + \sum_{i>r} u_i H_{i,r}(\mathbf{u})/H_{r,r}(\mathbf{u})\right),$$

and $v_i := u_i$ for $i \neq r$. Then the Jacobian $\left|\frac{\partial \mathbf{v}}{\partial \mathbf{u}}(\mathbf{0})\right| \neq 0$, so $\mathbf{u} \mapsto \mathbf{v}$ is a change of coordinates in a neighbourhood $U'' \subset U'$ of $\mathbf{0}$ by the inverse function theorem. Furthermore,

$$f(\mathbf{v}) = \sum_{i\leq r} \pm v_i^2 + \sum_{i,j>r} v_i v_j H'_{i,j}(\underline{v})$$

for $\mathbf{v} \in U''$. □

Corollary 1.50 *A non-degenerate critical point is always isolated.*

Definition 1.51 A subset of a topological space is called *residual* if it is an intersection of countably many open dense subsets. We say that a property of elements of a topological space is *generic* if it holds for a residual subset.

Theorem 1.52 *Morse functions form a residual subset of* $C^\infty(M)$.

For a proof, see the book of Milnor [114]. Since every residual set is dense, every smooth manifold admits a Morse function, and every smooth function can be perturbed to a Morse function.

Now suppose that f is a Morse function such that different critical points have distinct values. This is a generic condition. If c is a regular value of f, then $f^{-1}(c)$ is a smooth submanifold of M by the implicit function theorem. This is the boundary of the *sub-level set*

$$M^c := f^{-1}((-\infty, c]).$$

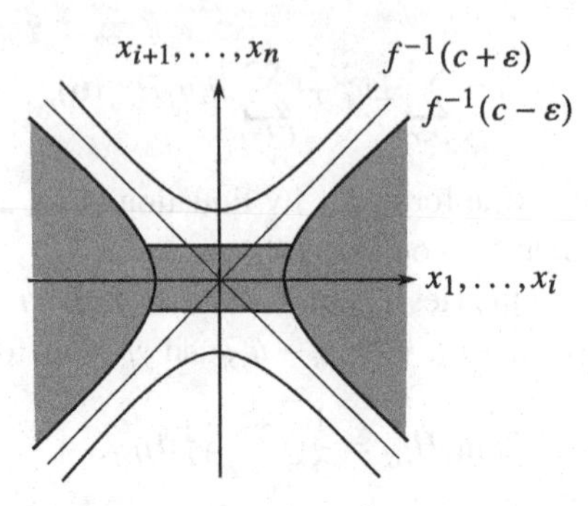

Figure 1.1 This figure shows the level sets $f^{-1}(c-\varepsilon)$, $f(c)$, and $f^{-1}(c+\varepsilon)$ of the function f in Lemma 1.49, where $c = f(\underline{0})$. The shaded region is $M^{c-\varepsilon}$ union the i-handle $D^i \times D^{n-i}$, which becomes diffeomorphic to $M^{c+\varepsilon}$ after smoothing the corners.

As c increases, both $f^{-1}(c)$ and M^c change smoothly as long as c does not pass a critical value. The key observation of Morse theory is that if c passes a point of $f(\mathrm{Crit}_i(f))$, the manifold M^c changes by attaching a thickened i-cell, called an n-dimensional i-handle. This is $D^i \times D^{n-i}$, which we attach along an embedding

$$h\colon \partial D^i \times D^{n-i} \hookrightarrow \partial M^c = f^{-1}(c)$$

and smooth the resulting corners. This follows from a simple local analysis of the normal form of f in Lemma 1.49. Indeed, if $c = f(\underline{0})$ and $\varepsilon > 0$ is small, then – up to diffeomorphism – we obtain $M^{c+\varepsilon}$ from $M^{c-\varepsilon}$ by gluing the i-handle that is the closure of

$$N(\{\, \underline{x} \in \mathbb{R}^n : x_{i+1} = \dots = x_n = 0 \,\}) \setminus M^{c-\varepsilon}$$

to $M^{c-\varepsilon}$; see Figure 1.1. We will study handle decompositions of manifolds in more detail in Section 1.6.

In particular, if we are only interested in the homotopy type of M^c, then it changes by attaching an i-cell. Successively attaching such cells, we obtain a CW complex homotopy equivalent to M. Indeed, by cellular approximation, we can homotope the attaching map of the i-cell into the $(i-1)$-skeleton of the complex corresponding to $M^{c-\varepsilon}$. Hence, using cellular homology, we obtain upper bounds on the Betti numbers of M: If $c_i = |\mathrm{Crit}_i(f)|$, then

$$b_i(M) \le c_i$$

for $i \in \{0, \dots, n\}$. These are called the *weak Morse inequalities*. Furthermore,

$$\chi(M) = \sum_{i=0}^{n} (-1)^n c_i.$$

Definition 1.53 A *Riemannian metric* g on a smooth manifold M is a positive definite, symmetric bilinear form g_p on T_pM for each $p \in M$ that varies smoothly in the sense that, for every coordinate chart $\phi\colon U \to \mathbb{R}^n$ and i, $j \in \{1,\dots,n\}$, the functions $g_p(\partial/\partial x_i, \partial/\partial x_j)$ are smooth in $p \in U$.

If we are given a Morse function f and a Riemannian metric g on M, then we can also compute the homology of M using f and g, as follows. Let v be the *gradient* of f with respect to the metric g, which is defined by the equation $g(v,w) = wf$ for all vector fields w on M. If c, $c' \in \mathbb{R}$ satisfy $c < c'$ and $f(\mathrm{Crit}(f)) \cap [c,c'] = \emptyset$, then the flow of v/vf gives a diffeomorphism between $f^{-1}([c,c'])$ and $f^{-1}(c) \times I$.

If p is an index i critical point of f, then choose local coordinates $(x_1,\dots,x_n)$ about p as in Lemma 1.49. Assume that g is the usual Euclidean metric in these coordinates; that is, $g(\partial/\partial x_i, \partial/\partial x_j) = \delta_{i,j}$. Then, in this chart,

$$v = 2(-x_1,\dots,-x_i,x_{i+1},\dots,x_n).$$

We see that the *stable manifold* $D_s(v,p)$ of v at p – the set of points of M on flow lines of v converging to p – is an i-disk. Similarly, the *unstable manifold* $D_u(v,p)$ of v at p is a disk of dimension $n-i$. If $c = f(p)$, the manifold $M^{c+\varepsilon}$ is obtained from $M^{c-\varepsilon}$ by attaching a neighbourhood of the i-disk $D_s(v,p) \setminus M^{c-\varepsilon}$.

If g is sufficiently generic, the unstable manifold $D_u(v,p)$ and the stable manifold $D_s(v,q)$ are transverse for any p, $q \in \mathrm{Crit}(f)$. This is called the *Morse–Smale condition* for the pair (f,g). Then, for $i \in \mathbb{Z}$, we defined the chain group $C_i(f,g)$ to be the free Abelian group generated by $\mathrm{Crit}_i(f)$.

Remark 1.54 It is often more convenient to work with *gradient-like vector fields* instead of Riemannian metrics. Given a Morse function f on a smooth n-manifold M, a gradient-like vector field for f is a vector field v such that $vf(p) > 0$ whenever $p \in M \setminus \mathrm{Crit}(f)$; furthermore, there are local coordinates $(x_1,\dots,x_n)$ about each critical point p in which

$$f(x_1,\dots,x_n) = f(p) - x_1^2 - \cdots - x_i^2 + x_{i+1}^2 + \cdots + x_n^2,$$

and the vector field is the Euclidean gradient:

$$v = 2(-x_1,-\dots,-x_i,x_{i+1},\dots,x_n).$$

The advantage of working with gradient-like vector fields is that they are easier to manipulate and deform than Riemannian metrics.

To define the *boundary map*

$$\partial_i : C_i(f,g) \to C_{i-1}(f,g),$$

it is sufficient to give the matrix A of ∂_i with respect to the bases $\mathrm{Crit}_i(f)$ and $\mathrm{Crit}_{i-1}(f)$. For $p \in \mathrm{Crit}_i(f)$ and $q \in \mathrm{Crit}_{i-1}(f)$, the entry $a_{q,p}$ of the matrix A is the signed count of flow lines of $-v$ from p to q. The union of these flow lines is precisely $D_s(v,p) \cap D_u(v,q)$, which are transverse, and hence is a manifold of dimension

$$i + (n - i + 1) - n = 1.$$

As each flow line is one-dimensional, if we quotient $D_s(v,p) \cap D_u(v,q)$ by the reparametrisation $\mathbb{R}$-action along the flow lines, we obtain a zero-manifold. We call this the *moduli space* of flow lines of $-v$ from p to q. We can identify this zero-manifold with

$$D_s(v,p) \cap D_u(v,q) \cap f^{-1}(c)$$

for $c \in (f(q), f(p))$ if $f(q) < f(p)$, and is empty otherwise. This zero-manifold is compact, as it is a transverse intersection of two spheres in $f^{-1}(c)$, and hence there are finitely many flow lines of $-v$ from p to q. The disks $D_s(v,p)$ and $D_u(v,q)$ are oriented by the local coordinates chosen about p and q, and hence each component of their intersection carries an orientation. A flow line from p to q is *positive* if it is oriented from p to q, and is *negative* otherwise.

To see that the graded free Abelian group

$$C_*(f,g) := \bigoplus_{i \in \mathbb{N}} C_i(f,g)$$

with the endomorphism $\partial_* := \bigoplus_{i \in \mathbb{N}} \partial_i$ is a chain complex, we need to check that $\partial_{i-1} \circ \partial_i = 0$. Let $p \in \mathrm{Crit}_i(f)$ and $r \in \mathrm{Crit}_{i-2}(f)$. Then the coefficient of r in $\partial_{i-1} \circ \partial_i(p)$ counts so-called *broken flow lines*, which go from p to a critical point $q \in \mathrm{Crit}_{i-1}(f)$, and then from q to r. This is exactly the boundary of $D_s(v,p) \cap D_u(v,r)$, which is the union of flow lines of $-v$ from p to r, and is a two-manifold. It is not compact but can be compactified with broken flow lines. So, after quotienting by the reparametrisation $\mathbb{R}$-action, the moduli space of flow lines from p to r is a compact one-manifold with boundary broken flow lines; see Figure 1.2. On the other hand, one can show that every broken flow line is the boundary of such a moduli space, which is called the *gluing lemma*. The signed count of endpoints of a compact one-manifold is zero, and hence so is the algebraic count of broken flow lines from p to r.

The reader familiar with cellular homology can easily verify that the preceding construction recovers the cellular homology of the CW complex

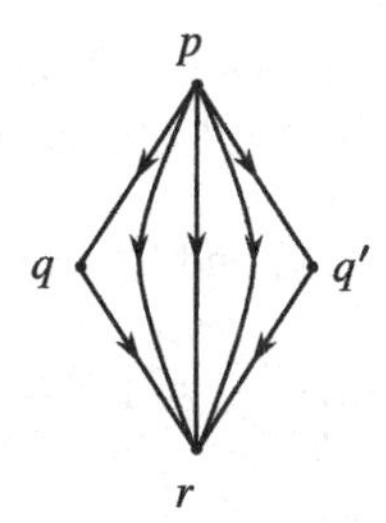

Figure 1.2 The moduli space of negative gradient flow lines between critical points p and r of index difference two is a one-manifold that can be compactified with broken flow lines.

corresponding to the Morse function f and metric g, which is homotopy equivalent to M. Hence

$$H_*(M) \cong H_*(C_*(f,g),\partial_*),$$

as claimed.

A simple consequence of the preceding description of the homology of a smooth manifold is *Poincaré duality*: If M is an n-manifold, then

$$H_*(M) \cong H^{n-*}(M).$$

Indeed, if f is a Morse function and g is a Riemannian metric on M that satisfy the Morse–Smale condition, then $-f$ is a Morse function on M with Morse complex

$$C_*(-f,g) \cong \mathrm{Hom}(C_*(f,g),\mathbb{Z}).$$

Furthermore, index i critical points of f correspond to index $n-i$ critical points of $-f$. In fact, the isomorphism of Poincaré duality is induced by cap product with the fundamental class of M; see, for example, Hatcher [57].

An extension of Morse homology to infinite-dimensional manifolds is called Floer homology, which we will study in Chapter 5. The key difficulty is that critical points no longer have finite index. Floer's observation was that the index difference of a pair of critical points can still be defined.

We conclude this section with the following application of Morse theory, due to Reeb:

Theorem 1.55 *Let M be a smooth, closed n-manifold that admits a Morse function f with only two critical points. Then M is homeomorphic to S^n.*

Proof Since f has only two critical points, one has to be the global minimum, which has index 0, and the other the global maximum, which has index n.

Hence f can be obtained by gluing an n-handle $h^n = D^n$ to a 0-handle $h^0 = D^n$ along a diffeomorphism $\phi\colon \partial h^n \to \partial h^0$. Consider the hemispheres

$$S^n_+ := S^n \cap \{x_{n+1} \geq 0\} \text{ and } S^n_- := S^n \cap \{x_{n+1} \leq 0\}$$

and the points $p_\pm := S^n \cap \{x_{n+1} = \pm 1\}$. We define the homeomorphism $H\colon M \to S^n$ as follows. For $x \in h^0$, let

$$H(x) = \left(x, -\sqrt{1-|x|^2}\right) \in S^n_-.$$

We extend this to h^n radially via

$$H(x) := \left(|x|\phi(x/|x|), \sqrt{1-|x|^2}\right) \in S^n_+. \qquad \square$$

Remark 1.56 First, note that the preceding homeomorphism H is not smooth at the centre of the n-handle. The gluing map of the two handles is determined by the flow of a gradient vector field v on M between the minimum and the maximum. The smooth structure on an n-manifold obtained by gluing two n-disks depends on the isotopy class of the gluing map. This gluing map is called the *clutching function*. Kervaire and Milnor showed that there are 28 different smooth structures on S^7, all obtained by gluing two copies of D^7. We will give an overview of their work in Section 2.8.

For a detailed account of Morse theory, see the excellent book of Milnor [114]. Morse homology is developed in the book of Schwarz [158]. In this section we just gave an outline of the construction, and have omitted several deep analytical proofs.

1.6 Handle Decompositions and Surgery

When studying homotopy types of topological spaces, CW complexes play a fundamental role. These are built from cells of varying dimensions. In order to study n-manifolds, we instead consider handle decompositions, which are obtained by thickening cells so that they all become n-dimensional.

Definition 1.57 An *n-dimensional k-handle* h^k is $D^k \times D^{n-k}$, which we can think of as a thickened k-cell. We call k the *index* of the handle. The *core* of h^k is $D^k \times \{0\}$, the *cocore* is $\{0\} \times D^{n-k}$, the *attaching sphere* is $A(h^k) := S^{k-1} \times \{0\}$, and the *belt sphere* is $B(h^k) := \{0\} \times S^{n-k-1}$.

Given an n-manifold M with boundary and an embedding

$$\varphi\colon S^{k-1} \times D^{n-k} \hookrightarrow \partial M,$$

called the *attaching map*, we can *attach* the k-handle $D^k \times D^{n-k}$ to M along φ as follows. We take the disjoint union $M \sqcup (D^k \times D^{n-k})$, identify $x \in S^{k-1} \times D^{n-k}$ with $\varphi(x) \in \partial M$, and round the corners. The result is unique up to diffeomorphism, and we denote it by $M(\varphi)$.

To specify φ, we need to define an embedding of the attaching sphere $S^{k-1} \times \{0\}$, together with a normal framing; that is, an identification of a neighbourhood of the image with $S^{k-1} \times D^{n-k}$. So, the normal bundle of the attaching sphere has to be trivial, and we also need to specify a trivialisation of the normal bundle (the framing).

When attaching a handle, ∂M changes by removing the image of φ, and gluing in $D^k \times S^{n-k-1}$ using $\varphi|_{S^{k-1}\times S^{n-k-1}}$. This leads to the notion of surgery.

Definition 1.58 Let X be a smooth n-manifold, and $S \subset X$ an embedded k-sphere with trivial normal bundle and normal framing

$$\nu\colon S^k \times D^{n-k} \to N(S).$$

Then the result of *surgery* on X along the framed sphere (S, ν) is

$$X(S, \nu) := (X \setminus N(S)) \cup_{\nu|_{S^k\times S^{n-k-1}}} (D^{k+1} \times S^{n-k-1}).$$

More generally, one could glue using any automorphism of $S^k \times S^{n-k-1}$, which will lead to the notion of Dehn surgery in dimension three. Sometimes we will only write $X(\nu)$ instead of $X(S, \nu)$.

If $\Phi \in \mathrm{Diff}(X)$ is an automorphism of X, then it induces a diffeomorphism

$$\Phi(\nu)\colon X(\nu) \to X(\Phi \circ \nu).$$

We saw in Section 1.5 that, when we pass a critical point of index i of a Morse function, the sub-level set changes by attaching an i-handle. The attaching map is given by the negative gradient flow from the critical point. Consequently, the level set changes by a surgery.

Recall that a CW complex is obtained by taking a collection of zero-cells (a discrete topological space), then attaching one-cells, followed by two-cells, and so on. The analogous construction for handles is called a handlebody.

Definition 1.59 An n-dimensional *handlebody* is obtained by taking finitely many n-dimensional zero-handles, and recursively attaching one-handles, followed by two-handles, and so on.

A *handle decomposition* of the smooth n-manifold M consists of a handlebody H, together with a diffeomorphism $\phi\colon H \to M$.

Analogously, we can define relative handle decompositions, built on an n-manifold with boundary. Furthermore, the resulting handlebody can be a manifold with boundary.

We now describe an operation on handle decompositions, obtained by isotoping the attaching map of one of the handles. While this is intuitively simple, the precise definition is notationally cumbersome. Suppose that we are given an n-dimensional handlebody H, obtained by consecutively attaching the handles $h^1, \dots, h^k$ to a manifold M with boundary, with attaching maps $\varphi_1, \dots, \varphi_k$, respectively. If we write

$$H_i = M(\varphi_1)\dots(\varphi_i),$$

then $H = H_k$. Given an isotopy $\{\varphi_i^t : t \in I\}$ such that $\varphi_i^0 = \varphi_i$, it can be extended to an ambient isotopy

$$\{G^t \in \mathrm{Diff}(\partial H_{i-1})\colon t \in I\}$$

using the isotopy extension theorem (Theorem 1.36), and the extension is unique up to isotopy. For every integer $j \in (i, k]$, the map G^1 induces a diffeomorphism

$$G_j' := G^1(\varphi_i)\dots(\varphi_j)\colon \partial H_j \to \partial H_j',$$

where $H_j' = H_i(\varphi_i')\dots(\varphi_j')$ for $\varphi_j' = G_{j-1}' \circ \varphi_j$.

Let $\partial H_{i-1} \times I$ be a collar neighbourhood of ∂H_{i-1}, such that $\partial H_{i-1} \times \{1\}$ is identified with ∂H_{i-1}. Choose a smooth, monotonically increasing function $f\colon \mathbb{R} \to I$ such that $f(t) = 0$ for $t \leq 0$ and $f(t) = 1$ for $t \geq 1$. Then the map

$$\psi_{i-1}(x, t) := \left(G^{f(t)}(x), f(t)\right)$$

for $(x, t) \in \partial H_{i-1} \times I$ can be extended to all of H_{i-1} as the identity. By construction, $\psi_{i-1}|_{\partial H_{i-1}} = G^1$, so ψ_{i-1} can further be extended to the handles $h^i, \dots, h^k$ to obtain a diffeomorphism $\psi\colon H_k \to H_k'$. If $\phi\colon H_k \to N$ is a handle decomposition of an n-manifold N, then the result of isotoping the attaching map φ_i of the handle h^i is defined to be $\phi \circ \psi^{-1}\colon H_k' \to N$.

Lemma 1.60 *If we attach a k-handle h^k followed by an l-handle h^l such that $k \geq l$, then we can isotope the attaching map of the l-handle to be disjoint from h^k.*

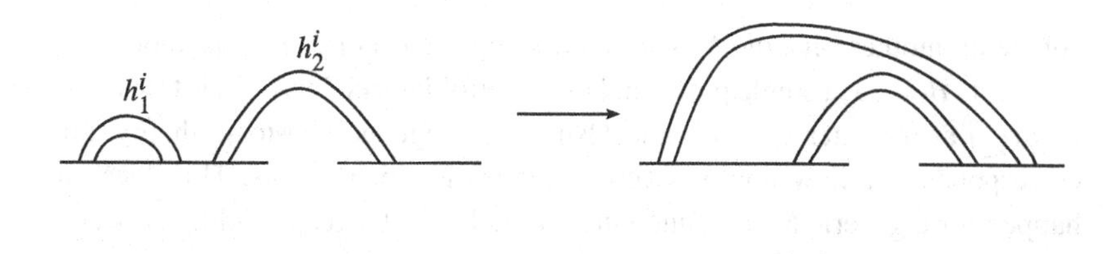

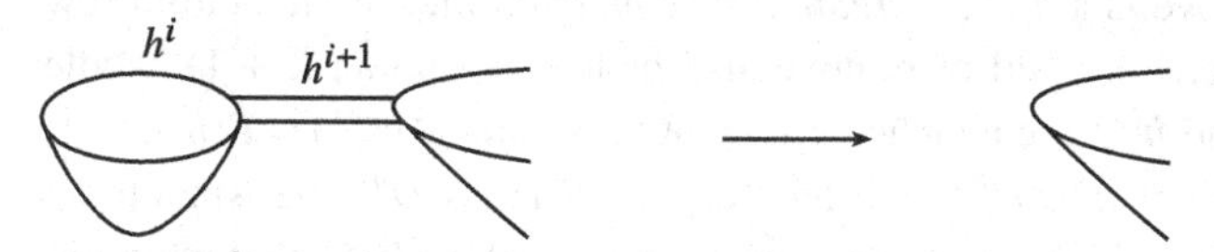

Figure 1.3 In the top row, the handle h_1^i slides over the handle h_2^i. In the bottom row, the handles h^i and h^{i+1} cancel.

Proof Note that $\dim(A(h^l)) = l - 1$, $\dim(B(h^k)) = n - k - 1$, and

$$(l - 1) + (n - k - 1) \leq n - 2.$$

So, we can perturb the attaching map of h^l such that its image is disjoint from $B(h^k)$. Then we can push off h^l from h^k by isotoping its attaching map radially along the rays of the core of h^k. □

Proposition 1.61 *Every closed smooth manifold admits a handle decomposition.*

Proof Let M be a closed smooth n-manifold. By Theorem 1.52, there is a Morse function $f: M \to \mathbb{R}$, and we can arrange that $f|_{\mathrm{Crit}(f)}$ is injective. Consider $M^c := f^{-1}((-\infty, c])$ for $c \in \mathbb{R}$. When $c < \min(f)$, we have $M^c = \emptyset$. We saw in Section 1.5 that, if $c \in f(\mathrm{Crit}_i(f))$ and $\varepsilon > 0$ is sufficiently small, then $M^{c+\varepsilon}$ is obtained from $M^{c-\varepsilon}$ by attaching an n-dimensional i-handle. It follows that M can be constructed by recursively attaching handles. To arrange that the handles are attached such that their indices are non-decreasing, we apply Lemma 1.60. □

Any two handle decompositions of a smooth manifold are related by a set of elementary moves. However, we have to pass through handle decompositions where the handles are not necessarily attached with indices in increasing order during the intermediate steps.

The first move is an isotopy of the attaching map of one of the handles. As defined earlier, we modify the attaching maps of subsequently attached handles suitably as well. A special case is a *handle slide*; see the top row of Figure 1.3. Here, one isotopes the attaching map of an i-handle h_1^i over the belt sphere of another i-handle h_2^i such that, along the isotopy, the image of $A(h_1^i)$ intersects $B(h_2^i)$ at a single point, and the manifold traced by $A(h_1^i)$ is transverse to $B(h_2^i)$ at the intersection point. Using the language of Morse theory, this corresponds to a flow line between two index i critical points. This does not happen for a generic Morse function and gradient-like vector field, but occurs in one-parameter families.

The second move is a *handle creation or cancellation*; see the bottom row of Figure 1.3. Here, we add or remove an i-handle h^i and an $(i+1)$-handle h^{i+1}, where h^i and h^{i+1} are attached consecutively and $|A(h^{i+1}) \cap B(h^i)| = 1$. This is possible since $h^i \cup h^{i+1}$ is diffeomorphic to a disk D^n, after smoothing corners. This is intuitively clear but requires a careful analysis that we omit; see Milnor [115].

Theorem 1.62 *Any two handle decompositions of a closed smooth manifold can be connected by isotopies of the attaching maps of the handles, and handle creations and cancellations.*

Proof To prove that the preceding two moves are sufficient to connect any two handle decompositions of the same manifold, we again use Morse theory. We first construct Morse functions f and f' and gradient-like vector fields v and v', respectively, such that (f, v) and (f', v') induce the two handle decompositions. We obtain these by gluing model functions and vector fields on each handle. We then choose a generic one-parameter family of smooth functions f_t for $t \in I$, such that $f_0 = f$ and $f_1 = f'$. The only singularity appearing in such a family is of the form

$$f_t(x_1, \dots, x_n) = c - x_1^2 - \cdots - x_i^2 + x_{i+1}^2 + \cdots + x_{n-1}^2 + x_n^3 \pm t x_n. \quad (1.2)$$

This is called a birth-death singularity and is the suspension of the family $x^3 \pm tx$ for $t \in I$. When the sign is positive, a pair of critical points die, and are born when the sign is negative. The critical points have indices i and $i+1$, and there is a single gradient flow line connecting them. So, if we choose v_t to be the Euclidean gradient of f_t in this local coordinate system, the attaching sphere of the higher index handle will intersect the belt sphere of the lower index one in a single point.

As t increases, the attaching spheres of the handles corresponding to the critical points of f_t change by isotopies, except when f_t has a birth-death singularity, as in Equation (1.2), which corresponds to a handle creation or cancellation. □

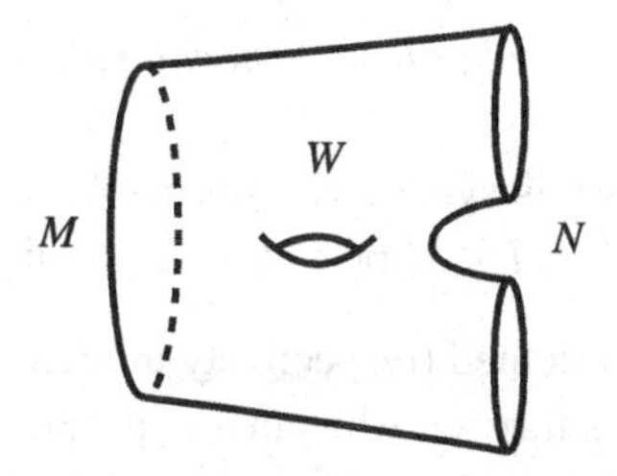

Figure 1.4 A cobordism W from M to N.

1.7 Cobordisms

Cobordism provides a coarser notion of equivalence between manifolds than homeomorphism or diffeomorphism. The idea stems from an attempt of Poincaré to define homology, was formalised by Pontryagin, and further studied by Thom.

Definition 1.63 Let M and N be closed n-manifolds. A *cobordism* W from M to N is a compact $(n+1)$-manifold with boundary such that

$$\partial W = M \sqcup N,$$

where $\sqcup$ denotes 'disjoint union'; see Figure 1.4. If M and N are oriented, we say that W is an *oriented cobordism* if W is also oriented and

$$\partial W = -M \sqcup N.$$

Two manifolds are called *cobordant* if there is a cobordism between them.

Remark 1.64 If we did not require W to be compact, every manifold M would be cobordant to $\varnothing$ via $M \times [0, \infty)$.

We can define cobordisms in both the topological and the smooth category. Given cobordisms W from M_0 to M_1 and W' from M_1 to M_2, we can define their composition $W' \circ W$ by gluing them along M_1. In the smooth category, we need to smooth the corners along M_1; however, the result is unique up to diffeomorphism fixing M_0 and M_2. Hence cobordism is a transitive relation.

If W is an oriented cobordism from M to N, then

$$\partial W = -M \sqcup N = -(-N) \sqcup -M,$$

so we can also view W as a cobordism from $-N$ to $-M$. We denote this by $\overline{W}$, and call it the *reverse* of W.

Proposition 1.65 *Oriented cobordism is an equivalence relation.*

Proof Oriented cobordism is symmetric, since $-W$ is a cobordism from N to M. It is reflexive, since $W = I \times M$ provides a cobordism from M to M. □

Cobordism classes of oriented (respectively unoriented) manifolds with the operation of disjoint union form an Abelian group, called the *oriented* (respectively *unoriented*) *cobordism group* of n-manifolds, which we denote by Ω_n^{SO} (respectively $\mathfrak{N}_n$). The identity is the class of the empty n-manifold, which consists of those n-manifolds that are boundaries of compact $(n+1)$-manifolds. In Ω_n^{SO}, the inverse of $[M]$ is $[-M]$, since we can view $I \times M$ as a cobordism from $M \sqcup -M$ to $\emptyset$. Similarly, in $\mathfrak{N}_n$, the inverse of $[M]$ is $[M]$. These groups have been determined using the pioneering work of René Thom [168]. For the purposes of low-dimensional topology, the following groups will be important:

$$\Omega_n^{SO} \cong \begin{cases} \mathbb{Z} & \text{if } n = 0, \\ 0 & \text{if } n = 1, \\ 0 & \text{if } n = 2, \\ 0 & \text{if } n = 3, \\ \mathbb{Z} & \text{if } n = 4. \end{cases} \tag{1.3}$$

Exercise 1.66 Prove Equation (1.3) and compute $\mathfrak{N}_n$ for $n \in \{0, 1, 2\}$.

More generally, Cartesian product of manifolds endows

$$\Omega_*^{SO} := \bigoplus_{i \geq 0} \Omega_i^{SO}$$

with a ring structure. Thom proved that $\Omega_*^{SO} \otimes \mathbb{Q}$ is a polynomial ring generated by the cobordism classes of the complex projective spaces $\mathbb{CP}^{2i}$ for $i > 0$. More generally, two oriented manifolds are oriented cobordant if and only if they have the same Stiefel–Whitney and Pontryagin numbers. (See Section A.3 for an overview of characteristic classes and Milnor–Stasheff [118] for the computation of Ω_*^{SO}.)

The signature is a numerical cobordism invariant that we now define. Let M be a closed oriented $4k$-manifold with fundamental class $[M] \in H_{4k}(M; \mathbb{R})$. The cup product defines a symmetric bilinear form

$$Q_M : H^{2k}(M; \mathbb{R}) \otimes H^{2k}(M; \mathbb{R}) \to \mathbb{Z}$$

by the formula $Q_M(x \otimes y) := \langle x \cup y, [M] \rangle$, which is non-degenerate by Poincaré duality. Then the *signature* $\sigma(M) \in \mathbb{Z}$ of M is the signature of Q_M; that is, the dimension of a maximal positive definite subspace minus the dimension of a maximal negative definite subspace of $H^{2k}(M;\mathbb{R})$. Thom showed that the signature is a cobordism invariant. In fact, it defines a homomorphism

$$\sigma\colon \Omega_{4k}^{SO} \to \mathbb{Z}.$$

The group $\Omega_4^{SO} \cong \mathbb{Z}$ is generated by $\mathbb{CP}^2$. Since $\sigma(\mathbb{CP}^2) = 1$, the map $\sigma\colon \Omega_4^{SO} \to \mathbb{Z}$ is an isomorphism.

Let B_k be the kth Bernoulli number. These are defined as the coefficients in the expansion

$$\frac{z}{e^z - 1} = 1 - \frac{z}{2} + \frac{B_1}{2!}z^2 - \frac{B_2}{4!}z^4 + \frac{B_3}{6!}z^6 \mp \dots;$$

see Milnor–Stasheff [118]. For example, $B_2 = 1/30$.

The signature of a smooth manifold can be expressed using Pontryagin numbers, due to the work of Hirzebruch and Thom:

Theorem 1.67 *Let M be a smooth, closed, and oriented n-manifold. If $n = 4$, then*

$$\sigma(M) = \frac{1}{3}p_1[M].$$

If $n = 8$, then

$$\sigma(M) = \frac{1}{45}(7p_2[M] - p_1^2[M]).$$

More generally, there is a k-variable polynomial $L_k(x_1, \dots, x_k)$ for every $k \in \mathbb{N}$ such that

$$\sigma(M^{4k}) = \langle\, L_k(p_1(M), \dots, p_k(M)), [M]\, \rangle.$$

If a monomial $x_1^{i_1} \dots x_k^{i_k}$ appears with non-zero coefficient in L_k, then $\sum_j j \cdot i_j = k$. The coefficient of x_k in L_k is $2^{2k}(2^{2k} - 1)B_k/(2k)!$.

To prove this theorem, it suffices to check it for the generators $\mathbb{CP}^{2i}$ of the oriented cobordism ring Ω_*^{SO}.

We say that two cobordisms from M_0 to M_1 are *equivalent* if they are diffeomorphic relative to $M_0 \sqcup M_1$. The *cobordism category* Cob_n has objects closed (smooth and/or orientable) n-manifolds and morphisms equivalence classes of cobordisms between them. The identity cobordism of M is given by $I \times M$. A *topological quantum field theory*, or TQFT in short, is a certain functor from Cob_n to the category of vector spaces and linear maps that takes disjoint unions to tensor products. To be completely rigorous, we should define cobordisms as follows:

Definition 1.68 A cobordism from M_0 to M_1 as a 5-tuple

$$(W, N_0, N_1, \phi_0, \phi_1),$$

where $\partial W = N_0 \sqcup N_1$ (or $-N_0 \sqcup N_1$ if we are working with oriented manifolds), and $\phi_i \colon N_i \to M_i$ are diffeomorphisms or homeomorphisms (depending on the category) for $i \in \{0, 1\}$.

Why this is necessary becomes clear when trying to construct the identity cobordism from M to M. Using the less rigorous definition, this would be a manifold W with $\partial W = M \sqcup M$, which is impossible unless $M = \emptyset$.

Definition 1.69 The identity cobordism from M to M is given by the tuple $(I \times M, \{0\} \times M, \{1\} \times M, e_0, e_1)$, where $e_i(i, x) = x$ for $x \in M$ and $i \in \{0, 1\}$.

More generally, Definition 1.68 makes it very easy to associate a cylindrical cobordism to a diffeomorphism:

Definition 1.70 The *cylindrical cobordism* W_ψ associated to a diffeomorphism $\psi \colon M \to N$ is the tuple

$$(I \times M, \{0\} \times M, \{1\} \times M, \phi_0, \phi_1),$$

where $\phi_0(0, x) = x$ and $\phi_1(1, x) = \psi(x)$ for $x \in M$.

Definition 1.71 We say that the cobordisms

$$(W, N_0, N_1, \phi_0, \phi_1) \text{ and } (W', N_0', N_1', \phi_0', \phi_1')$$

from M_0 to M_1 are *equivalent* if there is a diffeomorphism (or homeomorphism) $\Phi \colon W \to W'$ such that $\Phi|_{N_i} = (\phi_i')^{-1} \circ \phi_i$ for $i \in \{0, 1\}$.

Morphisms in the smooth cobordism category are equivalence classes of cobordisms, since the composition of two cobordisms is only well defined up to equivalence due to the smoothing involved. Throughout this work, we usually use the less rigorous definition of cobordism, as it is typically straightforward to make the arguments precise.

Exercise 1.72 Let $\psi, \psi' \colon M \to N$ be diffeomorphisms of n-manifolds. Show that the cobordisms W_ψ and $W_{\psi'}$ are equivalent if and only if ψ and ψ' are pseudoisotopic.

Definition 1.73 If W is a cobordism from M_0 to M_1, then we say that $f\colon W \to [a,b]$ is a *Morse function* if it has only non-degenerate critical points that all lie in $\mathrm{Int}(W)$, and $f^{-1}(a) = M_0$ and $f^{-1}(b) = M_1$.

We have the following analogue of Proposition 1.61 for cobordisms:

Proposition 1.74 *If W is a cobordism from M_0 to M_1, then it admits a relative handle decomposition by successively attaching handles to $I \times M_0$ along $\{1\} \times M_0$ with non-decreasing indices.*

Proof We obtain a handle decomposition of W relative to $I \times M_0$ from a Morse function on W. We can arrange that the handles are attached with non-decreasing indices by Lemma 1.60. □

Note that $-f$ is a Morse function on the reversed cobordism $\overline{W}$. It induces a handle decomposition of $\overline{W}$ relative to $I \times M_1$ whose cellular chain complex is dual to the chain complex corresponding to f. Hence

$$H_*(W, M_0) \cong H^*(W, M_1),$$

which is known as *Poincaré–Lefschetz duality*.

Definition 1.75 If S is an embedded k-sphere with normal framing ν in the n-manifold M, we define the *trace of the surgery* on M along (S,ν) to be the cobordism $W(S,\nu)$ from M to $M(S,\nu)$ obtained by attaching an $(n+1)$-dimensional $(k+1)$-handle to $I \times M$ along $\{1\} \times S$ using the framing $1 \times \nu$.

Traces of surgeries admit Morse functions with a single critical point. They are known as *elementary cobordisms*. Since every cobordism admits a Morse function, every cobordism is a product of elementary cobordisms.

2
Higher-Dimensional Manifolds

This chapter provides insight into the theory of high-dimensional smooth manifolds. We first present Milnor's celebrated construction of an exotic seven-sphere. We then overview the Whitney trick, which is a move aimed at eliminating intersection points of transverse submanifolds of complementary dimensions. It is the key component of the proof of the h-cobordism theorem, a fundamental result that underlies the classification of higher-dimensional manifolds using surgery theory. We illustrate the power of surgery theory by proving Kervaire and Milnor's classification of smooth structures on the seven-sphere; there are 28 up to diffeomorphism. We conclude with a summary of Cerf theory, which will later be used to prove Kirby's fundamental result on a calculus of diagrams representing four-manifolds.

2.1 Milnor's Exotic Seven-Sphere

In this section, we outline Milnor's construction [111] of the first exotic smooth structure from 1956; see also Milnor–Stasheff [118, pp. 243–248] for an alternate and more detailed construction. This is a smooth manifold M homeomorphic but not diffeomorphic to S^7. It is the boundary of the total space E of a D^4-bundle over S^4. Let S^4_+ and S^4_- denote the upper and lower hemispheres of S^4, respectively. Any D^4-bundle over S^4 is obtained by gluing the trivial D^4-bundles $S^4_+ \times D^4$ and $S^4_- \times D^4$ over S^3 using a map $f\colon S^3 \to \mathrm{SO}(4)$.

Let $\mathrm{Sp}(1)$ be the group of unit quaternions, which we identify with S^3. The map

$$p\colon \mathrm{Sp}(1) \times \mathrm{Sp}(1) \to \mathrm{SO}(4)$$

taking the pair of unit quaternions (q_1, q_2) to the isometry of $\mathbb{H} \cong \mathbb{R}^4$ given by $q \mapsto q_1 q q_2$ for $q \in \mathbb{H}$ is a 2-fold covering map, since $p(q_1, q_2) = p(-q_1, -q_2)$.

It follows that

$$\pi_3(\mathrm{SO}(4)) \cong \pi_3(S^3 \times S^3) \cong \mathbb{Z} \oplus \mathbb{Z}.$$

The pair $(i, j) \in \mathbb{Z} \oplus \mathbb{Z}$ is represented by the map

$$f\colon \mathrm{Sp}(1) \cong S^3 \to \mathrm{SO}(4)$$

given by $f(x)y = x^i y x^j$ for $x \in \mathrm{Sp}(1)$ and $y \in \mathbb{H}$.

Hence, for each pair of integers (i, j), we obtain a four-disk bundle over the four-sphere. The boundary M has the homotopy type of a seven-sphere if and only if $i + j = \pm 1$. Choose $i = 2$ and $j = -1$. Using the fact that M is an S^3-bundle over S^4 constructed using the preceding explicit map f, one can describe a smooth function on M with only two critical points. Hence, by Reeb's theorem (Theorem 1.55), M is *homeomorphic* to S^7.

Suppose, by contradiction, that M were *diffeomorphic* to S^7. Then we could attach D^8 to E along $M = \partial E$ to obtain a closed, smooth eight-manifold N. Since E deformation retracts onto S^4 (the zero-section of the D^4-bundle), the manifold N is homotopy equivalent to a CW complex consisting of one zero-cell, one four-cell, and one eight-cell. Hence $H^i(N) \cong \mathbb{Z}$ for $i \in \{0, 4, 8\}$ and $H^i(M) \cong 0$ otherwise. So $\sigma(N) = \pm 1$, and we can choose the orientation of N such that $\sigma(N) = 1$. By the Hirzebruch–Thom signature formula (Theorem 1.67), we have

$$1 = \sigma(N) = \frac{1}{45}(7p_2[N] - p_1^2[N]). \tag{2.1}$$

Since the restriction map $H^4(N) \to H^4(S^4)$ is an isomorphism, $p_1(N)$ is completely determined by $p_1(TN|_{S^4})$ using the naturality of characteristic classes. By Kervaire [79, lemma 1.1], we have $p_1(N) = 6\mathrm{PD}([S^4])$, and so $p_1^2[N] = 36$. Combining this with Equation (2.1), we obtain that

$$p_2[N] = \frac{45 + p_1^2[N]}{7} = \frac{81}{7},$$

which is not an integer, and hence a contradiction to the assumption that M is diffeomorphic to S^7. Furthermore, combined with Novikov's result [132] that the rational Pontryagin classes are topological invariants, we also conclude that N is a topological eight-manifold that admits no smooth structure. The first example of a triangulable manifold that admits no smooth structure was given by Kervaire [80] in dimension 10.

In Section 2.8, we will prove a result of Kervaire and Milnor that there are exactly 28 pairwise non-diffeomorphic smooth structures on S^7. In the following sections, we develop the necessary techniques.

2.2 The Whitney Trick

The Whitney trick plays a fundamental role in manifold topology. Its failure in lower dimensions is the main reason why the classification of manifolds in dimensions three and four is more difficult and has a different flavour than in higher dimensions. While this book focuses on low-dimensional topology, it is important to understand the reason behind this distinction. The Whitney trick is the key component of the proof of the h-cobordism theorem, which is the subject of the following section. However, it was originally used by Whitney [178] to prove that every n-manifold embeds into $\mathbb{R}^{2n}$.

Let A and B be smooth submanifolds of the n-manifold M, and write $a = \dim(A)$ and $b = \dim(B)$. If A and B are transverse and $a + b = n$, then the intersection is a discrete set of points. When A, B, and M are oriented, each intersection point $p \in A \cap B$ has a positive or negative sign, which we denote by $\operatorname{sgn}(p) \in \{\pm 1\}$. We have $\operatorname{sgn}(p) = +1$ if and only if an oriented basis of T_pA followed by an oriented basis of T_pB gives an oriented basis of

$$T_pM = T_pA \oplus T_pB.$$

Note that this also depends on the order of the submanifolds A and B. In $B \cap A$, the intersection signs are $(-1)^{\dim(A)\dim(B)}$ times those in $A \cap B$, which is the sign of the permutation swapping the bases of T_pA and T_pB.

If A and B are closed and oriented, they represent homology classes $[A]$, $[B] \in H_*(M)$. Their algebraic intersection is

$$\#(A \cap B) := \sum_{p \in A \cap B} \operatorname{sgn}(p) \in \mathbb{Z}.$$

If $\alpha, \beta \in H^*(M)$ are the Poincaré duals of $[A]$ and $[B]$, respectively, then this agrees with $\langle \alpha \cup \beta, [M] \rangle$. In particular, $\#(A \cap B)$ only depends on the homology classes that A and B represent. More generally, $\alpha \cup \beta$ is dual to $A \cap B$, even if we do not assume that $\dim(A) + \dim(B) = \dim(M)$.

Now let $f \colon M \looparrowright N$ be an immersion. Suppose that f is *self-transverse*; that is, if $p, q \in M$ are distinct points such that $f(p) = f(q)$, then

$$df(T_pM) + df(T_qM) = T_{f(p)}N.$$

If $\dim(N) = 2\dim(M)$, then generically all self-intersections of f are isolated double points (points with two pre-images). Even when M and N are oriented, the double points of f do not necessarily have canonical signs, since there is no canonical ordering of the pre-images p and q. Swapping p and q changes the signs by the factor $(-1)^{\dim(M)^2}$. Hence, when $\dim(M)$ is even, the sign does not depend on the order of p and q.

Proposition 2.1 (Whitney trick) *Let A and B be oriented, smooth submanifolds of the oriented n-manifold M that intersect transversely, and write $a = \dim(A)$ and $b = \dim(B)$. Suppose the following hold:*

(i) $a + b = n$,
(ii) $n \geq 5$, $a \geq 1$, *and* $b \geq 3$,
(iii) p, $q \in A \cap B$ *have opposite signs,*
(iv) *when a is 1 or 2, then the map $\pi_1(M \setminus B) \to \pi_1(M)$ is injective,*
(v) *there are embedded paths s_A and s_B from p to q in A and B, respectively, with interiors disjoint from $A \cap B$, such that $s_A\bar{s}_B$ is contractible in M.*

Then there is an isotopy $h_t \colon M \to M$ for $t \in I$ such that $h_0 = Id_M$ and

$$h_1(A) \cap B = A \cap B \setminus \{p, q\}.$$

In other words, we can eliminate intersection points of opposite signs using an ambient isotopy, under the preceding assumptions.

Proof Since the curve $C := s_A\bar{s}_B$ is null-homotopic in M, there is a map $h \colon D^2 \to M$ such that $h|_{\partial D^2} = C$. As $n \geq 5$, we can perturb the map h to be a smooth embedding by Theorem 1.43 and the remark following it, and to be transverse to A and B along $\text{Int}(D^2)$. When a, $b \geq 3$, the latter implies that $h(\text{Int}(D^2)) \cap (A \cup B) = \emptyset$, since both A and B have codimension at least three, while D^2 is two-dimensional. When a is one or two, condition (iv) implies that C is also contractible in $M \setminus B$, and hence we can choose h such that $h(\text{Int}(D^2)) \cap B = \emptyset$. As the codimension of A is still $b \geq 3$, we have $h(\text{Int}(D^2)) \cap A = \emptyset$ by transversality. We can also assume that $T_x h(D^2) \cap T_x A = T_x s_A$ for every $x \in s_A$ and $T_x h(D^2) \cap T_x B = T_x s_B$ for every $x \in s_B$.

Let $v_1, \ldots, v_{a-1}$ be linearly independent normal vector fields along s_A tangent to A; see the top of Figure 2.1. Then $v_i(p)$ and $v_i(q)$ are normal to both $h(D^2)$ and B for $i \in \{1, \ldots, a-1\}$. We extend these vector fields to s_B such that they are normal to both $h(D^2)$ and B. This is possible since the normal to $Th(D^2) + TB$ over s_B is a trivial rank $a-1$ vector bundle. Furthermore, the frames $v_1(p), \ldots, v_{a-1}(p)$ and $v_1(q), \ldots, v_{a-1}(q)$ lie in the same path component of $\text{GL}(a-1, \mathbb{R})$ as $\text{sgn}(p) = -\text{sgn}(q)$.

Since D^2 is contractible, the normal bundle of $h(D^2)$ in M is a trivial $(n-2)$-bundle ε^{n-2}. The vector fields $v_1, \ldots, v_{a-1}$ give a map φ from the curve C to the Stiefel manifold $V_{a-1}(\mathbb{R}^{n-2})$, which is $(b-2)$-connected. (The Stiefel manifold $V_k(\mathbb{R}^l)$ consists of k-frames in $\mathbb{R}^l$, and $\pi_i(V_k(\mathbb{R}^l)) = 0$ for $i < l - k$.) Hence, the frame $v_1, \ldots, v_{a-1}$ extends to the normal bundle of $h(D^2)$ in M, giving a

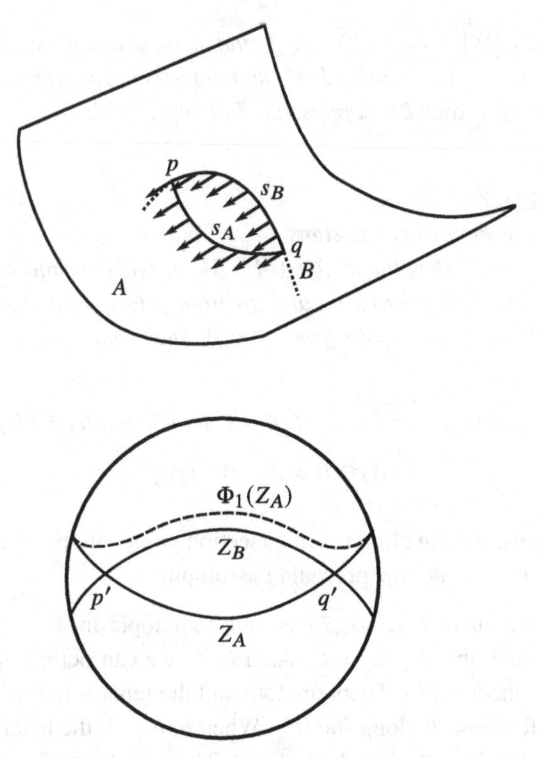

Figure 2.1 An illustration of the proof of the Whitney trick. The top shows the trivialisation of the normal bundle of the Whitney disk. The bottom shows the standard model in the disk.

splitting $\varepsilon^{n-2} \cong \varepsilon^{a-1} \oplus \varepsilon^{b-1}$. (The complement of ε^{a-1} is trivial since every bundle is trivial over D^2.)

We now consider the standard model $D^2 \times \mathbb{R}^{a-1} \times \mathbb{R}^{b-1}$, together with properly embedded arcs $Z_A, Z_B \subset D^2$ that intersect transversely in a pair of points p' and q'. See the bottom of Figure 2.1. Let D' be the closure of the component of $D^2 \setminus (Z_A \cup Z_B)$ disjoint from ∂D^2. The preceding splitting of the normal bundle of $h(D^2)$ allows us to define a diffeomorphism

$$\psi\colon D^2 \times \mathbb{R}^{a-1} \times \mathbb{R}^{b-1} \to N(h(D^2)),$$

such that

- $\psi(D' \times \{0\} \times \{0\}) = h(D^2)$,
- $\psi(p', 0, 0) = p$ and $\psi(q', 0, 0) = q$,
- $s_A \subset \psi(Z_A \times \{0\} \times \{0\})$ and $s_B \subset \psi(Z_B \times \{0\} \times \{0\})$,
- $\psi(Z_A \times \mathbb{R}^{a-1} \times \{0\}) \subset A$ and $\psi(Z_B \times \{0\} \times \mathbb{R}^{b-1}) \subset B$.

Let Φ_t for $t \in I$ be an isotopy of D^2 such that $\Phi_0 = \mathrm{Id}_{D^2}$, $\Phi_1(Z_A) \cap Z_B = \emptyset$, and $\Phi_t|_{\partial D^2} = \mathrm{Id}_{\partial D^2}$ for every $t \in I$. (First define Φ_t on Z_A, then extend it to D^2 using Theorem 1.36.) Furthermore, let $\lambda \colon \mathbb{R} \to I$ be a smooth function such that $\lambda(0) = 1$ and $\lambda(t) = 0$ whenever $|t| \geq 1$. Then we define the desired isotopy h_t of M on $\psi(D^2 \times \{x\} \times \{y\})$ to be $\psi \circ \Phi_{t\lambda(|(x,y)|)} \circ \psi^{-1}$ for $x \in \mathbb{R}^{a-1}$ and $y \in \mathbb{R}^{b-1}$, and the identity outside $N(h(D^2))$. □

A similar result holds about removing double points of opposite signs of immersed half-dimensional submanifolds. This is the key to proving the Whitney embedding theorem:

Theorem 2.2 (Whitney embedding theorem) *Every closed n-manifold M can be embedded into* $\mathbb{R}^{2n}$ *for* $n > 0$.

Proof Every one-manifold embeds in $\mathbb{R}^2$. Every closed orientable surface embeds in $\mathbb{R}^3$. Since $\mathbb{RP}^2$ embeds in $\mathbb{R}^4$, all closed non-orientable surfaces embed in $\mathbb{R}^4$ as well.

Now suppose that $n > 2$. Every n-manifold M can be immersed into $\mathbb{R}^{2n}$. Let $f \colon M \looparrowright \mathbb{R}^{2n}$ be an immersion with only transverse double points. Let p be a double point of f. We can locally modify f such that a single new double point q appears next to p. This is constructed by taking the connected sum of f with a small immersed S^n near p with a single double point, which is given by the following result:

Lemma 2.3 *There is an immersion* $S^n \looparrowright \mathbb{R}^{2n}$ *with exactly one double point. When n is even, we can specify the sign of the double point.*

Proof Let $J = [-1, 1]$. We will recursively construct immersions

$$j_n \colon J^n \looparrowright J^{2n}$$

that are standard along the boundary and have one double point. The case $n = 1$ is straightforward. Suppose we have constructed j_n, which has one double point with pre-images $x_-, x_+ \in J^n$. To obtain j_{n+1} from j_n, let $s_t \colon J^n \to J$ for $t \in J$ be a smooth one-parameter family of smooth functions such that $s_t|_{\partial J^n} \equiv 0$, $s_t(x_-) = -t/2$, and $s_t(x_+) = t/2$. Then we set

$$j_{n+1}(x,t) = (j_n(x), s_t(x), t)$$

for $x \in J^n$ and $t \in J$. We can reverse the sign of the double point when n is even by swapping the labels x_- and x_+ in the preceding construction. □

Call the modified immersion with the additional double point g, and let $g^{-1}(p) = \{p_A, p_B\}$ and $g^{-1}(q) = \{q_A, q_B\}$. Let s_A and s_B be disjoint, embedded paths in M connecting p_A with q_A and p_B with q_B, respectively. Let $h\colon D^2 \hookrightarrow \mathbb{R}^{2n}$ be an embedded Whitney disk bounding $g(s_A\bar{s}_B)$. Set $A := g(N(s_A))$ and $B := g(N(s_B))$; these are diffeomorphic to B^n. We can construct g such that p and q appear with opposite signs for any orientation of A and B. By the Whitney trick, there is an isotopy h_t for $t \in I$ of $\mathbb{R}^{2n}$ supported in $N(h(D^2))$ such that $h_0 = \mathrm{Id}_{\mathbb{R}^{2n}}$ and $h_1(A) \cap B = (A \cap B) \setminus \{p, q\}$. Then the map

$$f_1(x) := \begin{cases} h_1 \circ g(x) & \text{if } x \in N(s_A), \\ g(x) & \text{otherwise} \end{cases}$$

is an immersion of M into $\mathbb{R}^{2n}$ with one fewer double points than f. Repeating this procedure, we end up with an embedding of M into $\mathbb{R}^{2n}$. □

Whitney also showed that every n-manifold can be immersed into $\mathbb{R}^{2n-1}$. However, a generic map from an n-manifold into a $(2n-1)$-manifold is not an immersion, but has so-called Whitney umbrella singularities. These are end-points of double point curves of the immersion. We omit the proof, as we will not use the techniques involved in the rest of this book.

2.3 The h-Cobordism Theorem

The h-cobordism theorem is the key technical tool that allows one to reduce the classification of simply connected smooth and topological manifolds in dimension at least five to algebraic topology, proven by Smale [163]. It was extended to simply connected topological manifolds in dimension four by Freedman [37]. However, Donaldson [29] showed it fails in the smooth category in dimension four. In dimension three, it is still open whether it holds, and is equivalent to the smooth four-dimensional Poincaré conjecture. In dimension two, it follows from the classical three-dimensional Poincaré conjecture, proven by Perelman.

Definition 2.4 Let W be a cobordism from M_0 to M_1. Then we say that W is an *h-cobordism* (where 'h' stands for 'homotopy') if the embeddings $e_i\colon M_i \hookrightarrow W$ are homotopy equivalences for $i \in \{0, 1\}$.

Remark 2.5 By the relative Hurewicz theorem, when W, M_0, and M_1 are simply connected, the following are equivalent:

(i) the embedding e_0 is a homotopy equivalence,
(ii) $H_*(W, M_0) = 0$,

(iii) $H^*(W, M_1) = 0$,
(iv) $H_*(W, M_1) = 0$,
(v) e_1 is a homotopy equivalence.

Exercise 2.6 Prove the claim in Remark 2.5.

We now state and prove the h-cobordism theorem in dimensions at least 6.

Theorem 2.7 (h-cobordism theorem) *If $n \geq 6$ and W is a simply connected h-cobordism between the $(n-1)$-manifolds M_0 and M_1, then W is diffeomorphic to $I \times M_0$. In particular, M_0 and M_1 are diffeomorphic.*

Proof We outline the proof in the smooth category, using handle decompositions. The proof is described using the equivalent language of Morse functions and gradient-like vector fields in the beautiful book of Milnor [115].

By Proposition 1.74, there is a relative handle decomposition of W built on $I \times M_0$ such that the handles are attached with non-decreasing indices. Our goal is to cancel all the handles, after which we end up with the cylindrical cobordism $I \times M_0$. Recall that, whenever the attaching sphere of an $(i+1)$-handle h^{i+1} intersects the belt sphere of an i-handle h^i transversely in a single point, then we can cancel h^i and h^{i+1}.

Step 1. First, we cancel the zero-handles and n-handles. By assumption, both M_0 and W are path-connected. Attaching a zero-handle increases the number of components by one. Hence, for each zero-handle $h^0 \approx D^n$, there is a one-handle $h^1 \approx D^1 \times D^{n-1}$ such that $|A(h^1) \cap B(h^0)| = 1$, where $A(h^1) = S^0 \times \{0\}$ and $B(h^0) = S^{n-1}$. Then $h^0 \cup h^1 \approx D^n$, and we can cancel h^0 and h^1. By reversing the cobordism W, we can similarly cancel all the n-handles against some $(n-1)$-handles.

Step 2. Next, we remove all one-handles and $(n-1)$-handles. This is the most difficult step of the proof. Let N be the outgoing end of the cobordism W_2 obtained from $I \times M_0$ by attaching all the one-handles and two-handles of W. We claim that $\pi_1(N) = 1$. Indeed, we can construct W from $I \times N$ by attaching the handles of index at least three of W to $\{1\} \times N$, and the duals of the one-handles and two-handles of W to $\{0\} \times N$. These duals have indices $n-1$ and $n-2$, respectively. Note that $n-2 > 3$. If h is a handle of index $i \geq 3$ attached to a manifold P, then $\pi_1(P \cup h) \cong \pi_1(P)$. Indeed, $P \cap h \cong S^{i-1}$, so $\pi_1(P \cap h) = 1$. As $\pi_1(h) = 1$, the Seifert–van Kampen theorem implies that $\pi_1(P \cup h) \cong \pi_1(P)$. It follows that

$$\pi_1(N) \cong \pi_1(W) \cong 1.$$

We now attach a cancelling pair of two-handle h^2 and three-handle h^3 to W_2 along N, then attach the remaining handles of W. Let $h^1 \approx D^1 \times D^{n-1}$ be a one-handle of W. Since there are no zero-handles, h^1 is attached along $\{1\} \times M_0$, which is path-connected. For some $x \in \partial D^{n-1}$, connect the endpoints of the arc $D^1 \times \{x\} \subset \partial h^1$ in $\{1\} \times M_0$ disjoint from the attaching spheres of the one-handles and two-handles of W. This is possible since $\dim(N) \geq 3$. Let S be the resulting closed curve in N.

Since $\pi_1(N) = 1$, there is a homotopy of $A(h^2)$ to S. This homotopy can be viewed as a continuous map from $S^1 \times I$ to N. As $\dim(N) \geq 5$, this map can be perturbed to a smooth embedding, giving rise to an isotopy of $A(h^2)$ to S. Modify the handle decomposition of W using this isotopy. Since now $|A(h^2) \cap B(h^1)| = 1$, we can cancel h^1 and h^2. Hence, we have traded the one-handle h^1 with the three-handle h^3.

If we reverse W, handles of index $n-1$ become one-handles, so we can trade these for $(n-3)$-handles. Repeating this process, we can get rid of all one-handles and $(n-1)$-handles.

Step 3. Finally, we modify the handle decomposition of W using handle slides and isotopies such that we end up with pairs of neighbouring index handles (h^i, h^{i-1}) such that $|A(h^i) \cap B(h^{i-1})| = 1$, after which we can cancel h^i and h^{i-1}.

From the handle decomposition, we obtain a relative CW complex homotopy equivalent to (W, M_0) consisting of the cores of the handles. This gives rise to the cellular chain complex $C_*(W, M_0)$:

$$0 \longrightarrow C_{n-2} \longrightarrow \cdots \longrightarrow C_i \xrightarrow{\partial_i} C_{i-1} \longrightarrow \cdots \longrightarrow C_2 \longrightarrow 0;$$

see Section 1.5. Recall that C_i is the free Abelian group generated by i-handles with oriented cores, and the coefficient of the basis vector h^{i-1} in $\partial_i h^i$ for an i-handle h^i is the algebraic intersection number of $B(h^{i-1})$ and $A(h^i)$. (This corresponds to the algebraic number of flow lines between the corresponding critical points of a Morse function and Morse–Smale gradient-like vector field inducing the handle decomposition, which appears in the definition of Morse homology.)

Step 3a. We first recursively construct bases

$$\mathcal{B}_i = (z_1^i, \ldots, z_{n_i}^i, b_1^i, \ldots, b_{n_{i-1}}^i)$$

of C_i for $i \in \{2, \ldots, n-2\}$ such that $n_1 = n_{n-2} = 0$ and

$$\partial_i b_k^i = z_k^{i-1} \tag{2.2}$$

for $i \in \{3,\ldots,n-2\}$ and $k \in \{1,\ldots,n_{i-1}\}$. Let $\mathcal{B}_2 = (z_1^2,\ldots,z_{n_2}^2)$ be an arbitrary basis of C_2. We choose $b_1^3,\ldots,b_{n_2}^3 \in C_3$ such that they satisfy Equation (2.2). This is possible since $H_2(W, M_0) = 0$. We then extend this to a basis $(z_1^3,\ldots,z_{n_3}^3, b_1^3,\ldots,b_{n_2}^3)$ of C_3. We can continue this procedure since $H_i(W, M_0) = 0$ for all i; see Remark 2.5.

Step 3b. Given a finitely generated free Abelian group, we can connect any two bases using the following elementary operations:

(i) replace b_i with $b_i + b_j$ for some $i \neq j$,
(ii) replace b_i with $-b_i$,
(iii) permute the basis elements.

In other words, any invertible integral matrix can be diagonalised using elementary row operations (we are only allowed to multiply rows by -1, and not by arbitrary integers). Hence, we can connect the basis of $C_*(W, M_0)$ given by the handles to the algebraically constructed basis in Step 3a using the preceding elementary operations. Furthermore, each elementary operation can be realised by an operation on handles. Namely, move (i) corresponds to a handle slide, move (ii) to reversing the orientation of the core of the handle, and move (iii) to relabelling the handles.

Step 3c. Once we have a handle decomposition of W relative to $I \times M_0$ that gives a basis of $C_*(W, M_0)$ satisfying Equation (2.2), let h^i be a handle representing b_k^i and h^{i-1} a handle representing z_k^{i-1} for some $i \in \{3,\ldots,n-2\}$. The attaching sphere $A \approx S^{i-1}$ of h^i intersects the belt sphere $B \approx S^{n-i}$ of h^{i-1} *algebraically* once. Let N be the outgoing end of the cobordism obtained by attaching all handles of W of index at most $i-1$ to $I \times M_0$. As in Step 2, one can use the Seifert–van Kampen theorem to show that $\pi_1(N) = 1$. Note that $\dim(N) \geq 5$. By possibly reversing W, we can assume that $\dim(A) = i - 1 \geq 2$ and $\dim(B) = n - i \geq 3$.

Now suppose that $\dim(A) = 2$; that is, $i = 3$. Let N' be the outgoing end of the cobordism to which we attached h^{i-1}, and call its attaching sphere A'. Then $\pi_1(N') = 1$, and $N' \setminus A' \approx N \setminus B$, since N is obtained from N' by surgery along A'; see Definition 1.58. As $\dim(N') \geq 5$ and $\dim(A') = 1$, removing A' from N' does not change its fundamental group since any homotopy connecting two loops in $N' \setminus A'$ is generically disjoint from A'.

So the map $\pi_1(N \setminus B) \to \pi_1(N)$ is injective when $\dim(A) = 2$. Hence, we can apply the Whitney trick (Proposition 2.1) to $A, B \subset N$ to cancel all but one intersection points in $A \cap B$ in positive–negative pairs by isotoping the attaching map of h^i, and finally cancelling h^i and h^{i-1} once $|A \cap B| = 1$. This way, we can cancel all the handles of W in pairs, leaving us with $I \times M_0$. □

Using the terminology of Definition 1.70, the conclusion of the h-cobordism theorem is that W is equivalent to a cylindrical cobordism W_ψ for some diffeomorphism $\psi: M_0 \to M_1$.

The simple-connectivity assumption is necessary: Milnor showed that the seven-manifolds $L(7,1) \times S^4$ and $L(7,2) \times S^4$ with fundamental group $\mathbb{Z}_7$ are h-cobordant but not diffeomorphic. Here $L(7,1)$ and $L(7,2)$ are three-dimensional lens spaces, which we will study in detail in Section 3.7.

Donaldson gave an example of a five-dimensional smooth h-cobordism W from M_0 to M_1 that is not diffeomorphic to $I \times M_0$. On the other hand, Freedman proved the Whitney trick in the topologically locally flat category in dimension four, which implies the h-cobordism theorem for topological five-dimensional h-cobordisms, and gave his classification of simply connected topological four-manifolds that we will review in Section 6.2.

As an application of the h-cobordism theorem, we first give a characterisation of the smooth n-disc for $n \geq 6$. This plays an important role when classifying exotic spheres.

Proposition 2.8 *Let W be a compact, simply connected, smooth n-manifold for $n \geq 6$ with simply connected boundary. Then the following are equivalent:*

(i) *W^n is diffeomorphic to D^n.*
(ii) *W^n is homeomorphic to D^n.*
(iii) *W^n is contractible.*
(iv) *W^n has the same integral homology as a point.*

Proof As the four statements are respectively weaker, it suffices to prove that (iv) implies (i). Let $D_0 \subseteq \mathrm{Int}(W)$ be a smooth n-disk. By excision and the long exact sequence of the pair (W, D_0),

$$H_*(W \setminus \mathrm{Int}(D_0), \partial D_0) \cong H_*(W, D_0) = 0.$$

Since $\pi_1(W \setminus \mathrm{Int}(D_0)) = 1$ and $\pi_1(\partial W) = 1$, the manifold $W \setminus \mathrm{Int}(D_0)$ is a simply connected h-cobordism from $\partial D_0 \approx S^{n-1}$ to ∂W by Remark 2.5. Hence, by the h-cobordism theorem, $W \setminus \mathrm{Int}(D_0)$ is diffeomorphic to the product $I \times \partial D_0$, and so W is diffeomorphic to D^n. □

We now state and prove the Generalised Poincaré Conjecture in dimensions at least six.

Theorem 2.9 (Generalised Poincaré Conjecture) *Let M be a closed, simply connected, smooth n-manifold that has the same homology as S^n, and suppose that $n \geq 6$. Then M is* homeomorphic *to S^n.*

Proof Choose a handle decomposition of M. Analogously to Step 1 of the proof of the h-cobordism theorem, we can arrange that there is a single zero-handle h^0. Then

$$W := M \setminus \mathrm{Int}(h^0)$$

is simply connected and has the same homology as a point by excision and the long exact sequence of the pair (M, W). Hence W is diffeomorphic to D^n by Proposition 2.8. Reattaching h^0 gives a twisted sphere, which is homeomorphic to S^n by Theorem 1.55, as it admits a Morse function with only two critical points. □

Note that the Generalised Poincaré Conjecture holds in every dimension in the topological category, but it is beyond the scope of this book to prove it in full generality. The one- and two-dimensional cases simply follow from the classification of manifolds in these dimensions. The three-dimensional case is the classical Poincaré conjecture, proven by Perelman using the Ricci flow. The four-dimensional case is due to Freedman. In dimension five, the result follows from work of Kervaire and Milnor, which in fact implies the result in the smooth category in dimensions five and six. However, as mentioned earlier, there are 28 non-diffeomorphic smooth structures on S^7. It is still unknown if there is an exotic smooth structure on S^4.

To perform the Whitney trick, one does not need to assume that the ambient manifold is simply connected, only that the pair of intersection points to be cancelled can be connected by paths in the submanifolds A and B such that they together form a contractible loop; see condition (v) of Proposition 2.1. This allowed independently Barden [8], Mazur [109], and Stallings [164] to extend the h-cobordism theorem to manifolds that are not simply connected, leading to the s-cobordism theorem. The necessary condition involves keeping track of how the handles are attached with respect to the fundamental group of the cobordism, and is phrased in terms of vanishing of the Whitehead torsion of the pair (W, M_0). This, in turn, is equivalent to the embedding $e_0 \colon M_0 \hookrightarrow W$ being a simple homotopy equivalence (hence the 's' in s-cobordism). We now introduce the relevant notions.

2.4 Torsion

Torsions play an important role in topology, and we will encounter variants of it when discussing three-manifolds. They are all based on the following algebraic construction, which we review from the book of Turaev [173].

Let R be a commutative ring, and let V be a finitely generated free R-module. If $\mathbf{b} = (b_1, \ldots, b_k)$ and $\mathbf{c} = (c_1, \ldots, c_k)$ are bases of V, then we write $[\mathbf{b}/\mathbf{c}] \in R^*$

for the determinant of the change-of-basis matrix $(\mathbf{b}/\mathbf{c}) := (a_{ij})_{i,j=1,\dots,k}$ such that

$$b_i = \sum_{j=1}^{k} a_{ij} c_j$$

for $i \in \{1, \dots, k\}$.

Let

$$C = (0 \longrightarrow C_m \xrightarrow{\partial_m} \cdots \xrightarrow{\partial_1} C_0 \longrightarrow 0)$$

be an *acyclic* chain complex over R; that is, $H_*(C) = 0$. Furthermore, let $\mathbf{c}_i$ be a basis of C_i for $i \in \{0, \dots, m\}$. Let $B_i = \operatorname{Im}(\partial_{i+1})$ be the boundaries in C_i. Since $H_i(C) = 0$, we have $B_i = \operatorname{Ker}(\partial_i)$. Let $\mathbf{b}_i$ be a basis of B_i for $i \in \{-1, \dots, m\}$, where $\mathbf{b}_{-1} = \mathbf{b}_m = \emptyset$. Finally, let $\mathbf{b}'_{i-1}$ be a lift of $\mathbf{b}_{i-1}$ under ∂_i. Then $\mathbf{b}_i\mathbf{b}'_{i-1}$ is a basis of C_i. Compare this with the situation in Step 3 of the proof of the h-cobordism theorem, where the basis c_i is given by the i-handles.

Definition 2.10 The *torsion* of the based, acyclic chain complex C is

$$\tau(C) := \prod_{i=0}^{m} [\mathbf{b}_i\mathbf{b}'_{i-1}/\mathbf{c}_i]^{(-1)^{i+1}} \in R^*.$$

Exercise 2.11 Show that $\tau(C)$ is independent of the choice of $\mathbf{b}_i$ and their lifts $\mathbf{b}'_i$. If C' is the acyclic chain complex C based by $\mathbf{c}'_0, \dots, \mathbf{c}'_m$, then

$$\tau(C') = \tau(C) \prod_{i=0}^{m} [\mathbf{c}_i/\mathbf{c}'_i]^{(-1)^{i+1}}.$$

For the s-cobordism theorem, we need to work with chain complexes over the group ring $\mathbb{Z}[\pi_1(W)]$, and so we need a more general construction called Whitehead torsion. Let R be a unital ring, and $\mathrm{GL}(n, R)$ invertible $n \times n$ matrices over R. There is a natural embedding $\mathrm{GL}(n, R) \hookrightarrow \mathrm{GL}(n+1, R)$ given by

$$A \mapsto \begin{pmatrix} A & 0 \\ 0 & 1 \end{pmatrix}.$$

We write $\mathrm{GL}(R)$ for the direct limit $\varinjlim \mathrm{GL}(n, R)$. The *first K-group* of R is obtained by abelianising $\mathrm{GL}(R)$:

$$K_1(R) := \mathrm{GL}(R)/[\mathrm{GL}(R), \mathrm{GL}(R)].$$

Note that the commutator subgroup $[\mathrm{GL}(R), \mathrm{GL}(R)]$ is generated by elementary matrices $I \pm aE_{ij}$ for $a \in R$ and $i, j \in \{1, \dots, n\}$. Hence elements of $K_1(R)$ are matrices over R up to elementary row operations. Every element of $K_1(\mathbb{Z})$

can be represented by a diagonal matrix diag($\pm 1, \ldots, \pm 1$), which we used in Step 3b of the proof of the h-cobordism theorem, but this is not the case over general rings. Furthermore, in $K_1(R)$, we have

$$A \cdot B = \begin{pmatrix} A & 0 \\ 0 & B \end{pmatrix}. \tag{2.3}$$

As $K_1(R)$ is Abelian, we can also write the group operation additively.

We now consider the case when R is the group ring $\mathbb{Z}[G]$ of a group G. Then

$$H_1(G) := G/[G, G]$$

embeds into $K_1(\mathbb{Z}[G])$ by viewing $g \in G$ as a 1×1 matrix (g). By Equation (2.3), this image consists of matrices of the form diag$(g_1, \ldots, g_n)$ for $g_i \in G$.

Definition 2.12 The *Whitehead group* of the group G is

$$\mathrm{Wh}(G) := K_1(\mathbb{Z}[G])/\pm H_1(G).$$

Let Y be a finite connected CW complex, and let X be a subcomplex of Y such that the embedding $e\colon X \hookrightarrow Y$ is a homotopy equivalence. In particular, $\pi_1(X) \cong \pi_1(Y)$; write $\pi := \pi_1(X)$. Let $p\colon \tilde{Y} \to Y$ be the universal covering of Y, and write $\tilde{X} := p^{-1}(X)$. Then $\tilde{X}$ is a deformation retract of $\tilde{Y}$, hence the chain complex $C_*(\tilde{Y}, \tilde{X})$ over $\mathbb{Z}[\pi]$ is acyclic, where the π acts by covering transformations. If we choose lifts of the i-cells in Y, we obtain a basis $\mathbf{c}_i$ of $C_i(\tilde{Y}, \tilde{X})$. These lifts are unique modulo the π-action. Choose a basis $\mathbf{b}_i$ of the boundaries B_i in $C_i(\tilde{Y}, \tilde{X})$ and a lift $\mathbf{b}_i'$ of $\mathbf{b}_i$ as in Definition 2.10.

Definition 2.13 Let X and Y be as in the preceding paragraph. Then the *Whitehead torsion* of the pair (Y, X) is

$$\tau(Y, X) := \sum_{i=0}^{m} (-1)^{i+1} (\mathbf{b}_i \mathbf{b}_{i-1}' / \mathbf{c}_i) \in \mathrm{Wh}(\pi).$$

This is independent of the choice of lifts of the cells in Y. For further details, see Turaev [173].

2.5 Simple Homotopy Equivalence

Definition 2.14 Let X be a finite CW complex. We write X^i for the i-skeleton of X. Assume that the complex Y is obtained from X by attaching a k-cell e^k along a map $\partial D^k \to X^{k-1}$, and a $(k+1)$-cell e^{k+1} along a map

$$g\colon \partial D^{k+1} \to X^k \cup e^k$$

such that $g^{-1}(e^k)$ is an open k-ball in ∂D^{k+1} mapped by g homeomorphically onto e^k. Then the inclusion $X \hookrightarrow Y$ is called an *elementary expansion*, and its homotopy inverse $Y \to X$ is called an *elementary collapse*.

Elementary expansions and collapses are analogous to handle creations and cancellations, where we think of handles as thickened cells.

Definition 2.15 A homotopy equivalence between finite CW complexes is called *simple* if it can be written as a composition of elementary expansions and collapses.

We can characterise simple homotopy equivalences using Whitehead torsion. For this, we first need to define the Whitehead torsion of a homotopy equivalence. Let $f\colon X \to Y$ be a cellular map between finite CW complexes that is a homotopy equivalence. Replacing Y with the mapping cylinder, we can homotope f to an embedding. The *mapping cylinder* of f is

$$M_f := ((X \times I) \sqcup Y)/_{(x,1)\sim f(x)}.$$

Note that $r\colon M_f \to Y$ given by $r(x,t) = f(x)$ for $(x,t) \in X \times I$ and $r|_Y = \mathrm{Id}_Y$ is a deformation retraction. Consider the embedding $e_X\colon X \hookrightarrow M_f$ given by $e_X(x) = (x,0)$ and the natural embedding $e_Y\colon Y \to M_f$. Then e_X is homotopic to $e_Y \circ f$, so e_X is a homotopy equivalence if and only if f is. Furthermore, M_f has a natural CW structure such that $X \times \{0\}$ is a subcomplex and e_Y is a simple homotopy equivalence.

Definition 2.16 If $f\colon X \to Y$ is a homotopy equivalence, as earlier, we define its *Whitehead torsion* as

$$\tau(f) := \tau(M_f, X \times \{0\}).$$

In particular, if X is a subcomplex of Y and f is the embedding, then $\tau(f) = \tau(Y,X)$.

Theorem 2.17 *A cellular homotopy equivalence $f\colon X \to Y$ between finite CW complexes is simple if and only if $\tau(f) = 0$.*

For a proof, see (22.2) in Cohen [24]. It is similar but simpler than the proof of the s-cobordism theorem that we present in the next section, so we omit it. In some texts, a homotopy equivalence f is defined to be simple if $\tau(f) = 0$, and it is a consequence of the preceding theorem that it can be written as a composition of elementary expansions and collapses.

2.6 The s-Cobordism Theorem

Definition 2.18 A cobordism W from M_0 to M_1 is an *s-cobordism* if it is an h-cobordism and $\tau(W, M_0) = 0$.

We are now ready to state the s-cobordism theorem:

Theorem 2.19 (s-cobordism theorem) *If W is a connected s-cobordism from M_0 to M_1 such that $\dim(W) \geq 6$, then W is diffeomorphic to $I \times M_0$. In particular, M_0 and M_1 are diffeomorphic.*

Furthermore, if $\pi = \pi_1(M_0)$, then the map $W \mapsto \tau(W, M_0)$ is a bijection between diffeomorphism classes relative to M_0 of h-cobordisms from M_0 and $Wh(\pi)$.

Proof We follow the exposition of Kervaire [81].

Proposition 2.20 *Let W be an h-cobordism from M to M', such that M and M' are connected and oriented, and $\dim(W) = n + 1 \geq 6$. Then, for every integer $r \in [2, n-2]$, we can write W in the form*

$$W \approx (I \times M) \cup h_1^r \cup \cdots \cup h_\alpha^r \cup h_1^{r+1} \cup \cdots \cup h_\alpha^{r+1}, \tag{2.4}$$

where $h_1^r, \ldots, h_\alpha^r$ are r-handles and $h_1^{r+1}, \ldots, h_\alpha^{r+1}$ are $(r+1)$-handles.

Proof Choose a handle decomposition of W. Recall that in Step 2 of the proof of the h-cobordism theorem, we traded one-handles for three-handles. Motivated by this, we successively eliminate handles of index $q = 0, 1, \ldots, r-1$ by trading them for $(q+2)$-handles. We then reverse W and apply the same procedure to eliminate handles of indices $n+1, n, \ldots, r+2$.

We denote the i-handles by $h_1^i, \ldots, h_{\alpha_i}^i$ and the attaching map of h_j^i by φ_j^i. We write X_q for $I \times M$ with h_j^i for $i \leq q$ attached, and $Y_q := \partial X_q \setminus (\{0\} \times M)$ for the outgoing end of X_q. Furthermore, let

$$Y_q^0 := Y_q \setminus \bigcup_i \varphi_i^{q+1}(S^q \times B^{n-q}).$$

Then $Y_q^0 \subset Y_q$ and $Y_q^0 \subset Y_{q+1}$.

We can eliminate the zero-handles and n-handles of W as in Step 1 of the proof of the h-cobordism theorem.

We trade one-handles with three-handles as follows: The cobordism X_1 is the boundary connected sum of $I \times M$ and α_1 solid tori $S^1 \times D^n$. Hence

$$\pi_1(X_1) = \pi_1(M) * \langle x_1, \ldots, x_{\alpha_1} \rangle,$$

where x_i is represented by the core of the one-handle h_i^1. Let

$$\langle y_1, \ldots, y_u \mid r_1, \ldots, r_v \rangle$$

be a presentation of $\pi_1(M)$. We write ρ_i for the conjugacy class in $\pi_1(X_1)$ of the attaching map of h_i^2 for $i \in \{1,\dots,\alpha_2\}$. Then

$$\pi_1(X_2) = \langle\, y_1,\dots,y_u,x_1,\dots,x_{\alpha_1} \,|\, r_1,\dots,r_v,\rho_1,\dots,\rho_{\alpha_2}\, \rangle.$$

Since the embedding $M \hookrightarrow X_2$ induces an isomorphism on π_1, we can write x_i in $\pi_1(X_2)$ as a word w_i in $y_1,\dots,y_u$. The embedding $Y_1^0 \hookrightarrow Y_1$ induces an isomorphism on π_1. Suppose that $\xi_i \in \pi_1(Y_1^0)$ maps to $x_i w_i^{-1}$ in $\pi_1(Y_1) \cong \pi_1(X_1)$. Then ξ_i maps to 1 in $\pi_1(Y_2) \cong \pi_1(X_2)$. Let $\psi_i' \colon S^1 \times D^{n-1} \to Y_1^0$ be an embedding that represents ξ_i and intersects the belt circle of h_i^1 transversely once, but is disjoint from the belt circles of the other one-handles. Since $\dim(Y_2) \geq 5$, we can isotope ψ_i' to an embedding $\psi_i \colon S^1 \times D^{n-1} \hookrightarrow Y_2$ that lies in an $\mathbb{R}^n$ chart. We attach a two-handle g_i^2 to X_2 along each ψ_i, followed by a three-handle g_i^3 that cancels g_i^2. We then isotope the attaching map ψ_i of g_i^2 to ψ_i', and cancel it with h_i^1. This way, we have traded h_i^1 with g_i^3.

Now suppose that we have eliminated all handles of index less than q for some $q \in [2,r)$. Let a_i^s be a lift of the core of the handle h_i^s to the universal cover $\widetilde{X}_s$ of X_s. Let $\widetilde{W}$ denote the universal cover of the cobordism W, and let $\widetilde{X}$ be the universal cover of $I \times M$. Then the homology group $H_s(\widetilde{X}_s,\widetilde{X}_{s-1})$ is freely generated by $a_1^s,\dots,a_{\alpha_s}^s$ over $\mathbb{Z}[\pi]$. (We can identify this with the twisted cellular chain group $C_s(W,M;\mathbb{Z}[\pi])$.) We denote by

$$\partial \colon H_{s+1}(X_{s+1},X_s) \to H_s(X_s,X_{s-1})$$

the boundary map.

Let $f \colon S^q \hookrightarrow Y_q$ be an embedding. A lift of f to $\widetilde{X}_q$ represents a homology class of the form $[\tilde{f}] = \sum_i x_i a_i^q \in H_q(\widetilde{X}_q,\widetilde{X})$, where the $x_i \in \mathbb{Z}[\pi]$ are independent of the choice of lift $\tilde{f}$ up to overall multiplication by an element $\gamma \in \pi$. The following result will be used to show that if two handles cancel algebraically, then they can be isotoped to cancel geometrically; see Step 3c of the proof of the h-cobordism theorem.

Lemma 2.21 *The embedding $f \colon S^q \hookrightarrow Y_q$ for $q \in [2,n-2)$ is isotopic to $g \colon S^q \hookrightarrow Y_q$ such that*

(i) *$g(S^q)$ intersects the belt sphere $B(h_i^q)$ transversely in a single point, and*
(ii) *$g(S^q) \cap B(h_j^q) = \emptyset$ for $j \neq i$*

if and only if $[\tilde{f}] = \pm\gamma a_i^q \in H_q(\widetilde{X}_q,\widetilde{X})$ for some $\gamma \in \pi$.

Proof An isotopy of f in Y_q does not change the homology class of its lift. If $f(S^q)$ intersects $B(h_i^q)$ transversely in a single point and $f(S^q) \cap B(h_j^q) = \emptyset$ for $j \neq i$, it is clear that $[\tilde{f}] = \pm\gamma a_i^q$ for some $\gamma \in \pi$.

Conversely, suppose that $[\tilde{f}] = \pm\gamma a_i^q$. We can arrange that $f(S^q)$ intersects $B(h_k^q)$ transversely for $k \in \{1,\ldots,\alpha_q\}$. Choose a basepoint

$$p \in Y_q \setminus \left(f(S^q) \cup \bigcup_k B(h_k^q) \right)$$

and a path η_k from p to each $B(h_k^q)$ that induces the lift a_k^q of the core of h_k^q and a path η from p to $f(S^q)$ that induces the lift $\tilde{f}$ of f. If $Q \in f(S^q) \cap B(h_k^q)$ for some k, we choose a path η_Q that goes from $\eta(1)$ to Q in $f(S^q)$ and then from Q to $\eta_k(1)$ in $B(h_k^q)$. Finally, let $\gamma_Q := \eta\eta_Q\overline{\eta}_k$, which is a loop based at p. Let ε_Q be the intersection sign of $f(S^q)$ and $B(h_k^q)$ at Q. Then

$$[\tilde{f}] = \sum_{k=1}^{\alpha_q} \sum_{Q \in f(S^q) \cap B(h_k^q)} \varepsilon_Q \gamma_Q a_k^q.$$

As $[\tilde{f}] = \pm\gamma a_i^q$, we have

$$\sum_{Q \in f(S^q) \cap B(h_k^q)} \varepsilon_Q \gamma_Q = \pm\delta_{i,k}\gamma$$

for every $k \in \{1,\ldots,\alpha_q\}$. Hence, we can pair the points of $f(S^q) \cap B(h_k^q)$ for $k \neq i$ into pairs (Q, Q') such that $\gamma_Q = \gamma_{Q'}$ and $\varepsilon_Q = -\varepsilon_{Q'}$. For $k = i$, we can do the same leaving a single intersection point. We can then eliminate Q and Q' using the Whitney trick: If we connect Q to Q' on $f(S^q)$ and Q' to Q on $B(h_k^q)$ with paths, the resulting loop ℓ is freely homotopic to $\gamma_{Q'}\gamma_Q^{-1}$, which is contractible in Y_q and hence also in $Y_q \setminus \bigcup_k B(h_k^q)$, as the map

$$\pi_1\left(Y_q \setminus \bigcup_k B(h_k^q) \right) \longrightarrow \pi_1(Y_q)$$

is an isomorphism. So ℓ bounds a Whitney disk whose interior is disjoint from $f(S^q)$ since $q + 2 < n$. Removing all such pairs of intersection points using the Whitney trick, we are left with the configuration claimed in the statement of the lemma. □

Lemma 2.22 *Let $f\colon S^q \hookrightarrow Y_q^0$ be an embedding, and let $x_1,\ldots,x_\beta \in \mathbb{Z}[\pi]$ be arbitrary elements. Then there exists an embedding $g\colon S^q \hookrightarrow Y_q^0$ such that the compositions of f and g with the embedding $Y_q^0 \hookrightarrow Y_q$ are isotopic, and such that*

$$[\tilde{g}] = [\tilde{f}] + \sum_j x_j \partial a_j^{q+1},$$

where $[\tilde{f}]$, $[\tilde{g}] \in H_q(\widetilde{X}_q, \widetilde{X})$ are homology classes of lifts of f and g, respectively.

Proof The map $s_j = \varphi_j^{q+1}|_{S^q \times \{y_0\}}$ for $y_0 \in \partial D^{n-q-1}$ is isotopic in Y_{q+1} to a standard embedding of S^q in a neighbourhood of a point. For any $\gamma \in \pi$, we can take the connected sum of f with $\pm s_j$ along a suitable path such that it lifts to $\tilde{f} \pm \gamma \tilde{s}_j$. This connected sum is isotopic to f in Y_{q+1}. (This is an instance of a handle slide.) This way, we can change $\tilde{f}$ with any linear combination

$$\sum_j x_j \tilde{s}_j = \sum_j x_j \partial a_j^{q+1}$$

without changing the isotopy class of f in Y_{q+1}. □

We now return to the proof of eliminating handles of index q in Proposition 2.20. Since $H_*(W, M) = 0$, the boundary map

$$\partial \colon H_{q+1}(\widetilde{X}_{q+1}, \widetilde{X}_q) \to H_q(\widetilde{X}_q, \widetilde{X})$$

is surjective. For every $i \in \{1, \ldots, \alpha_q\}$, there are elements $x_{ij} \in \mathbb{Z}[\pi]$ such that $\sum_j x_{ij} \partial a_j^{q+1} = a_i^q$. We attach to X_{q+1} trivially $(q+1)$-handles $g_1^{q+1}, \ldots, g_{\alpha_q}^{q+1}$ along Y_q^0, followed by cancelling $(q+2)$-handles $g_1^{q+2}, \ldots, g_{\alpha_q}^{q+2}$. By Lemma 2.22, we can isotope the attaching map of g_i^{q+1} such that its lift represents the homology class $0 + \sum_j x_{ij} \partial a_j^{q+1} = a_i^q$ in $H_q(\widetilde{X}_q, \widetilde{X})$. We can then cancel g_i^{q+1} with h_i^q by Lemma 2.21. This way, we have traded h_i^q with g_i^{q+2}. This concludes the proof of Proposition 2.20. □

We now prove the first part of the s-cobordism theorem, so assume $\tau(W, M) = 0$. Suppose that we have a handle decomposition of W with only handles $h_1^r, \ldots, h_\alpha^r$ of index r and $h_1^{r+1}, \ldots, h_\alpha^{r+1}$ of index $r+1$, as in Proposition 2.20. Our goal is to cancel the r-handles and $(r+1)$-handles in pairs.

We can compute $\tau(W, M)$ from the complex

$$0 \longrightarrow H_{r+1}(\widetilde{X}_{r+1}, \widetilde{X}_r) \xrightarrow{\partial} H_r(\widetilde{X}_r, \widetilde{X}) \longrightarrow 0,$$

where $H_r(\widetilde{X}_r, \widetilde{X})$ and $H_{r+1}(\widetilde{X}_{r+1}, \widetilde{X}_r)$ are free $\mathbb{Z}[\pi]$-modules generated by $a_1^r, \ldots, a_\alpha^r$ and $a_1^{r+1}, \ldots, a_\alpha^{r+1}$, respectively. The class a_i^j is represented by the lift of the core of the handle h_i^j. Then

$$\partial a_i^{r+1} = \sum_j x_{ij} a_j^r$$

for an invertible matrix $A = (x_{ij})$ that represents $(-1)^{r+1}\tau(W, M)$ in the Whitehead group.

The assumption that $\tau(W, M) = 0$ means that we can transform x to the empty matrix using the following operations:

(i) Multiplying from the left a row of A with $\pm\gamma$ for $\gamma \in \pi$. This can be realised by changing the lift of the core of h_i^{r+1} and possibly reversing its orientation, which can be achieved using an isotopy of W.
(ii) Adding a row to another. This can be realised using Lemma 2.22.
(iii) Replacing $A \in \mathrm{GL}(n, \mathbb{Z}[\pi])$ with $A \oplus (1) \in \mathrm{GL}(n+1, \mathbb{Z}[\pi])$. This can be achieved by adding a cancelling pair of handles of indices r and $r+1$.
(iv) The reverse of operation (iii). This means that we have $\partial a_\beta^{r+1} = a_\beta^r$ for some β. By Lemma 2.21, we can isotope the attaching map of the handle h_β^{r+1} such that it cancels with h_β^r geometrically.

Using the preceding moves, we cancel all handles in pairs until we are left with $I \times M$.

We finally show the second part of the s-cobordism theorem. Given a closed, connected n-manifold M with $\pi = \pi_1(M)$, let $\tau_0 \in \mathrm{Wh}(\pi)$ be an element represented by an invertible matrix $A = (x_{ij})$ of size α. We attach to $I \times M$ along $\{1\} \times M$ the r-handles $h_1^r, \dots, h_\alpha^r$ in a trivial way for some $1 < r < n/2$. The resulting cobordism X_r has outgoing boundary

$$\partial_+ X_r := \partial X_r \setminus (\{0\} \times M) = (\{1\} \times M) \# (S^r \times S^{n-r}) \# \cdots \# (S^r \times S^{n-r}).$$

The embedding of $M \vee S^r \vee \cdots \vee S^r$ into $\partial_+ X_r$ induces an injective homomorphism of free $\mathbb{Z}[\pi]$-modules generated by $a_1^r, \dots, a_\alpha^r$ in $\pi_r(\partial_+ X_r)$. We now attach $(r+1)$-handles $h_1^{r+1}, \dots, h_\alpha^{r+1}$ with gluing maps

$$\varphi_i^{r+1} \colon S^r \times D^{n-r} \hookrightarrow \partial_+ X^r$$

represented by $\sum_j x_{ij} a_j^r \in \pi_r(\partial_+ X_r)$. This is possible by Theorem 1.43 since $r < n/2$. Then the resulting cobordism W is an h-cobordism between M and a manifold M' such that $\tau(W, M) = (-1)^{r+1} \tau_0$. □

Exercise 2.23 Show that the cobordism W constructed in the last step of the proof of the s-cobordism theorem is indeed an h-cobordism.

2.7 The Pontryagin Construction

Definition 2.24 We say that the n-manifold M is a *framed submanifold* of the $(n+k)$-manifold X if the normal bundle $\nu_{M \subseteq X}$ is trivial, and we are given an isomorphism

$$f \colon \nu_{M \subseteq X} \xrightarrow{\sim} \varepsilon^k.$$

This is equivalent to giving a diffeomorphism $N(M) \xrightarrow{\sim} M \times D^k$, or k linearly independent, ordered normal vectors at each point of M, up to homotopy.

We orient $\nu_{M \subset X}$ such that the trivialisation f is orientation-preserving. If X is oriented, then the framing induces an orientation of M since

$$\nu_{M \subset X} \oplus TM \cong TX|_M.$$

Definition 2.25 If (M_0, f_0) and (M_1, f_1) are closed, oriented, framed submanifolds of X, then we say that they are *framed cobordant* if there is a compact, oriented, framed submanifold (W, f) of $I \times X$ such that

$$\partial W = -(\{0\} \times M_0) \cup (\{1\} \times M_1),$$

and $f|_{\{i\} \times M_i} = f_i$ for $i \in \{0, 1\}$.

We denote by $\mathrm{Emb}^{fr}(n, k)$ the group of framed cobordism classes of closed, framed n-manifolds in $\mathbb{R}^{n+k}$, where the group operation is distant disjoint union of framed embeddings, and the inverse of a framed submanifold is its mirror image in a hyperplane. By the following result of Pontryagin, these groups agree with homotopy groups of spheres:

Theorem 2.26 *We have*

$$\mathit{Emb}^{fr}(n, k) \cong \pi_{n+k}(S^k).$$

Proof First, we define a map

$$\theta \colon \pi_{n+k}(S^k) \to \mathrm{Emb}^{fr}(n, k).$$

Every element of $\pi_{n+k}(S^k)$ can be represented by a *smooth* map

$$g \colon S^{n+k} \to S^k.$$

Let $p \in S^k$ be a regular value of g; that is, a point such that for every $x \in g^{-1}(\{p\})$ the differential $\mathrm{d}g_x \colon T_x S^{n+k} \to T_p S^k$ is surjective. Then $M := g^{-1}(\{p\})$ is a smooth submanifold of S^{n+k}.

Let $v_1, \dots, v_k$ be a basis of $T_p S^k$. For $x \in g^{-1}(\{p\})$, the differential $\mathrm{d}g_x$ induces an isomorphism

$$\overline{\mathrm{d}g}_x \colon T_x S^{n+k} / T_x M \xrightarrow{\sim} T_p S^k.$$

Then we can pull back $v_1, \dots, v_k$ along $\overline{\mathrm{d}g}_x$ to obtain a basis of the space $T_x S^{n+k} / T_x M$ for every $x \in M$. As

$$\nu_{M \subseteq S^{n+k}} = (TS^{n+k}|_M)/TM,$$

this is a framing f of M. (In other words, $\nu_{M \subseteq S^{n+k}}$ is the pullback $g^* \nu_{\{p\} \subseteq S^k} \cong \varepsilon^k$.) We define

$$\theta_p(g) := (M, f),$$

where we use that S^k is the one-point compactification of $\mathbb{R}^k$.

Let $R_g \subset S^k$ be the set of regular values of g. We claim that, if $p \in R_g$ and q is sufficiently close to p, then $q \in R_g$ and $\theta_p(g)$ and $\theta_q(g)$ are framed cobordant in $I \times S^{n+k}$. Indeed, since S^{n+k} is compact, the set R_g is open in S^k. Let $U_p \subset R_g$ be an open neighbourhood of p homeomorphic to B^k. For any $q \in U_p$, connect p and q with a path $\gamma\colon I \to U_p$. Then

$$\bigcup_{t \in I} (t, \theta_{\gamma(t)}(g))$$

gives the desired framed cobordism between $\theta_p(g)$ and $\theta_q(g)$.

To see that $\theta_p(g)$ is independent of the homotopy class of g up to framed cobordism, suppose that $g_0, g_1\colon S^{n+k} \to S^k$ are homotopic via a homotopy

$$G\colon I \times S^{n+k} \to S^k$$

and $p \in R_{g_0} \cap R_{g_1}$. Let $q \in R_{g_0} \cap R_{g_1}$ be a regular value of G that is so close to p that $\theta_p(g_i)$ and $\theta_q(g_i)$ are framed cobordant for $i \in \{0, 1\}$. The manifold $G^{-1}(\{q\}) \subset I \times S^{n+k}$ with the normal framing obtained by pulling back $v_1, \ldots, v_k$ along dG is a framed cobordism between $\theta_q(g_0)$ and $\theta_q(g_1)$. It follows that $\theta_p(g_0)$ and $\theta_p(g_1)$ are also framed cobordant.

Now suppose that $p, p' \in R_g$. Then choose a one-parameter family of rotations ψ_t of S^k for $t \in I$ such that $\psi_1(p) = p'$. Then p' is a regular value of $g' := \psi_1 \circ g$, and $\theta_{p'}(g') = \theta_p(g)$. Furthermore, g and g' are homotopic via $\psi_t \circ g$ and $p' \in R_g \cap R_{g'}$. So, $\theta_{p'}(g')$ and $\theta_{p'}(g)$ are framed cobordant by the previous paragraph. Hence, the map $\theta([g]) := [\theta_p(g)]$ is well defined.

We now define a map

$$\alpha\colon \mathrm{Emb}^{fr}(n, k) \to \pi_{n+k}(S^k).$$

Given $[(M, f)] \in \mathrm{Emb}^{fr}(n, k)$, the framing f can be viewed as a diffeomorphism $f\colon N(M) \to M \times D^k$. We define a map $g\colon S^{n+k} \to S^k$ on $N(M)$ as the composition

$$N(M) \xrightarrow{f} M \times D^k \xrightarrow{\text{proj}} D^k \longrightarrow D^k/\partial D^k \approx S^k,$$

while g maps $S^{n+k} \setminus N(M)$ to $[\partial D^k] \in D^k/\partial D^k \approx S^k$. Then we set

$$\alpha(M, f) := g.$$

To see that the homotopy class of $\alpha(M, f)$ is independent of the framed cobordism class of (M, f), we note that one can apply the same construction to a framed cobordism in $I \times S^{n+k}$ between (M_0, f_0) and (M_1, f_1) to obtain a homotopy between $\alpha(M_0, f_0)$ and $\alpha(M_1, f_1)$. Hence, we can write $\alpha([(M, f)]) := [g]$.

It is clear that $\theta \circ \alpha = \mathrm{Id}$, and we leave checking $\alpha \circ \theta = \mathrm{Id}$ as an exercise. □

Recall that the homotopy groups $\pi_m(S^n)$ vanish for $m < n$. Indeed, given a map $f: S^m \to S^n$, it is homotopic to a smooth map. By Sard's theorem (Theorem 1.42), the map f has a regular value $p \in S^n$, which means that $f^{-1}(\{p\}) = \emptyset$ since $m < n$. Then f is null-homotopic, since $S^n \setminus \{p\}$ is contractible. As an application of the Pontryagin construction, we compute further homotopy groups of spheres.

Theorem 2.27 *For every $n \geq 1$, we have $\pi_n(S^n) \cong \mathbb{Z}$.*

Proof By the Pontryagin construction,

$$\pi_n(S^n) \cong \mathrm{Emb}^{fr}(0, n),$$

where $\mathrm{Emb}^{fr}(0, n)$ is the group of framed cobordism classes of closed, framed zero-manifolds in $\mathbb{R}^n$. A closed, framed zero-manifold is simply a finite set of points $\mathbf{p} = \{p_1, \ldots, p_s\} \subset \mathbb{R}^n$, together with a normal framing $\varphi = (\varphi_1, \ldots, \varphi_s)$, where φ_i is an n-frame in $T_{p_i}\mathbb{R}^n$ for every $i \in \{1, \ldots, s\}$.

Each φ_i can be viewed as an $n \times n$ matrix. Then the map

$$\deg(\mathbf{p}, \varphi) := \sum_{i=1}^{s} \mathrm{sgn}(\det(\varphi_i))$$

descends to a homomorphism $\deg\colon \mathrm{Emb}^{fr}(0, n) \to \mathbb{Z}$. Indeed, a framed cobordism can be written as a product of elementary cobordisms, where each elementary cobordism cancels or creates a pair of points with normal framings of opposite signs. Furthermore, it is clearly additive under distant disjoint union. Note that deg is simply the degree of the map $S^n \to S^n$ corresponding to $(\mathbf{p}, \varphi)$ under the Pontryagin construction.

We now show that deg is an isomorphism. It is surjective since we can realise every integer $a \geq 0$ by taking a points with positive framings, and every $a < 0$ by choosing a points with negative framings. It is injective since $\deg(\mathbf{p}, \varphi) = 0$ implies that there are $s/2$ points with positive framing and $s/2$ points with negative framing, which we can cancel pairwise using elementary cobordisms. □

Proposition 2.28 *If $n \geq 3$, then $H_1(SO(n)) \cong \pi_1(SO(n)) \cong \mathbb{Z}_2$.*

Proof As SO(3) is homeomorphic to $\mathbb{RP}^3$, we have $\pi_1(\mathrm{SO}(3)) \cong \mathbb{Z}_2$. From the long exact sequence of the fibration

$$\mathrm{SO}(n) \to \mathrm{SO}(n+1) \to S^n$$

that assigns the first column vector to a matrix in $\mathrm{SO}(n+1)$ (see Theorem A.3), we obtain that $\pi_1(\mathrm{SO}(n)) \cong \pi_1(\mathrm{SO}(n+1))$ for $n \geq 3$. Hence $\pi_1(\mathrm{SO}(n)) \cong \mathbb{Z}_2$ for $n \geq 3$. But H_1 is the abelianisation of π_1, so $H_1(\mathrm{SO}(n)) \cong \mathbb{Z}_2$. □

Theorem 2.29 *For $n \geq 3$, we have $\pi_{n+1}(S^n) \cong \mathbb{Z}_2$.*

Proof By the Pontryagin construction, $\pi_{n+1}(S^n) \cong \mathrm{Emb}^{fr}(1,n)$. An element of $\mathrm{Emb}^{fr}(1,n)$ is represented by a closed, framed one-manifold (M,φ) in $\mathbb{R}^{n+1}$. Since $n+1 \geq 4$, any two embeddings of a one-manifold are ambient isotopic in $\mathbb{R}^{n+1}$. Hence, we can assume that $M \subset \mathbb{R}^2$. Then M has a canonical normal framing φ_0 obtained by extending a trivialisation of $\nu_{M \subset \mathbb{R}^2}$ with the standard basis vectors $e_3,\dots,e_{n+1}$. Comparing φ and φ_0 gives a map $f\colon M \to \mathrm{SO}(n)$.

We now define an isomorphism

$$h\colon \mathrm{Emb}^{fr}(1,n) \to \mathbb{Z}_2$$

by mapping (M,φ) to $[f(M)] \in H_1(\mathrm{SO}(n)) \cong \mathbb{Z}_2$; see Proposition 2.28. This is well defined since framed cobordant manifolds have homologous images. Indeed, suppose that the surface $S \subset I \times \mathbb{R}^{n+1}$ with normal framing ϕ is a cobordism from (M_0,φ_0) to (M_1,φ_1). There is a surface S' in $I \times \mathbb{R}^2$ with $\partial S = \{0\} \times M_0 \cup \{1\} \times M_1$ and homeomorphic to S relative to boundary. As $\dim(I \times \mathbb{R}^{n+1}) \geq 5$, we can isotope S to S'. Extend this to an ambient isotopy using the isotopy extension theorem. Let ϕ' be the normal framing of S' obtained by isotoping ϕ. The surface S' has a canonical normal framing ϕ_0 obtained by choosing a trivialisation of $\nu_{S' \subset I \times \mathbb{R}^2}$ and adding $e_3,\dots,e_{n+1}$. Comparing ϕ' and ϕ_0 gives the desired homology. The map h is a homomorphism, as it is additive under distant disjoint union.

To see that h is surjective, choose $M := S^1 \subset \mathbb{R}^2$ with normal framing ϕ obtained by twisting ϕ_0 by a generator of $\pi_1(\mathrm{SO}(n))$. Finally, to prove injectivity, suppose that (M,φ) is such that $f(M)$ is null-homologous in $\mathrm{SO}(n)$. Then there is a surface $S \subset I \times \mathbb{R}^2$ and a map $F\colon S \to \mathrm{SO}(n)$ such that $\partial S = \{0\} \times M$ and $F|_{\{0\}\times M} = f$. We now twist the standard normal framing ϕ_0 of S constructed above by F to obtain a framed cobordism from (M,φ) to $\emptyset$. □

Remark 2.30 Note that the condition $n \geq 3$ is necessary as $\pi_3(S^2) \cong \mathbb{Z}$. For an element (M,φ) of $\mathrm{Emb}^{fr}(1,2)$, normal framings φ are parametrised by $\pi_1(\mathrm{SO}(2)) \cong \mathbb{Z}$. Each component of $M \subset \mathbb{R}^3$ has a canonical framing by Remark 4.12, so we can define an isomorphism

$$\mathrm{Emb}^{fr}(1,2) \to H_1(\mathrm{SO}(2)) \cong \mathbb{Z}$$

as earlier. The generator of $\mathrm{Emb}^{fr}(1,2)$ is given by $M = S^1 \subset \mathbb{R}^2$, with normal framing φ rotating around once.

In fact, $\pi_3(S^2)$ is generated by the Hopf fibration $S^3 \to S^2$ with fibre S^1, defined by mapping $(z_1, z_2) \in S^3 \subset \mathbb{C}^2$ to $[z_1 : z_2] \in \mathbb{CP}^1 \approx S^2$. Consider the homotopy long exact sequence

$$\pi_n(S^1) \longrightarrow \pi_n(S^3) \longrightarrow \pi_n(S^2) \longrightarrow \pi_{n-1}(S^1)$$

of the Hopf fibration. As $\pi_k(S^1) = 0$ for $k > 1$, we see that $\pi_n(S^3) \cong \pi_n(S^2)$ for $n \geq 2$. In particular,

$$\pi_3(S^2) \cong \pi_3(S^3) \cong \mathbb{Z}$$

by Theorem 2.27, and

$$\pi_4(S^2) \cong \pi_4(S^3) \cong \mathbb{Z}_2$$

by Theorem 2.29.

The computation of $\pi_{n+2}(S^n)$ that we give in what follows can be viewed as the first instance of surgery theory, which we study in the next chapter. First, we need to introduce the Arf invariant.

Definition 2.31 A *quadratic form* over a field $\mathbb{F}$ is a map $q\colon V \to \mathbb{F}$, where V is a finite-dimensional $\mathbb{F}$-vector space, $q(\lambda v) = \lambda^2 q(v)$ for every $v \in V$ and $\lambda \in \mathbb{F}$, and

$$w(u, v) := q(u + v) - q(u) - q(v)$$

is bilinear.

Now suppose that $\mathbb{F} = \mathbb{F}_2$ and q is non-singular; that is, w is non-degenerate. Then w is a symplectic form, since w is symmetric and

$$w(v, v) = q(2v) - 2q(v) = 0.$$

Let $(a_1, b_1, \ldots, a_k, b_k)$ be a symplectic basis of V; that is, $w(a_i, a_j) = w(b_i, b_j) = 0$, and $w(a_i, b_j) = \delta_{ij}$ for $i, j \in \{1, \ldots, k\}$. Then the *Arf invariant* of q is defined as

$$\mathrm{Arf}(q) := \sum_{i=1}^{k} q(a_i)q(b_i) \in \mathbb{F}_2.$$

Note that, when $\mathrm{char}(\mathbb{F}) \neq 2$, the form w uniquely determines q, since $2q(v) = w(v, v)$. As we shall shortly see, this is not the case over $\mathbb{F}_2$. The following result is due to Arf.

Proposition 2.32 *Let V be an $\mathbb{F}_2$-vector space. The Arf invariant of a non-singular quadratic form $q\colon V \to \mathbb{F}_2$ is independent of the choice of symplectic basis of V, and it agrees with the value assumed most often by q (and is* *hence called a 'democratic invariant'). Two non-singular quadratic forms on V are equivalent if and only if they have the same Arf invariant.*

Proof We follow the exposition of Browder [14, III.1]. In essence, we classify all non-singular quadratic forms over $\mathbb{F}_2$ up to isomorphism.

Let (U, w) a two-dimensional symplectic vector space over $\mathbb{F}_2$ with symplectic basis (a, b). If a quadratic form q on U is compatible with w, then

$$q(a + b) + q(a) + q(b) = w(a, b) = 1,$$

so $|q^{-1}(1)| \in \{1, 3\}$. Since any two elements of $\{a, b, a + b\}$ form a symplectic basis, any two-dimensional non-singular quadratic form is isomorphic to either q_0 or q_1, defined by $q_0(a) = q_0(b) = 0$ and $q_1(a) = q_1(b) = 1$. Note that $q_0(a + b) = q_1(a + b) = 1$. Furthermore, q_0 and q_1 are not isomorphic as $|q_0^{-1}(1)| = 1$, while $|q_1^{-1}(1)| = 3$.

One can form the direct sum of quadratic forms. We claim that

$$(U \oplus U, q_0 \oplus q_0) \cong (U \oplus U, q_1 \oplus q_1). \tag{2.5}$$

Indeed, consider the symplectic bases

$$\begin{aligned}(a_1, b_1, a_2, b_2) &:= ((a, 0), (b, 0), (0, a), (0, b)) \text{ and} \\ (a_1', b_1', a_2', b_2') &:= ((a, a), (b, a), (a + b, a + b), (a + b, b)).\end{aligned}$$

Then $(q_0 \oplus q_0)(a_i) = (q_1 \oplus q_1)(a_i')$ and $(q_0 \oplus q_0)(b_i) = (q_1 \oplus q_1)(b_i')$.

Let $(a_1, b_1, \ldots, a_k, b_k)$ be a symplectic basis of V. Then $q|_{\langle a_i, b_i \rangle} \in \{q_0, q_1\}$ for $i \in \{1, \ldots, k\}$, and

$$q \in \{kq_0, (k-1)q_0 \oplus q_1\}$$

by Equation (2.5). Furthermore, it is immediate from the definition of the Arf invariant that $\mathrm{Arf}(kq_0) = 0$ and $\mathrm{Arf}((k-1)q_0 \oplus q_1) = 1$.

It remains to show that kq_0 maps the majority of the elements of V to zero and $(k-1)q_0 \oplus q_1$ maps the majority of the elements to one, which implies that these two forms are inequivalent. We proceed by induction on k. We have already checked the case $k = 1$. Set $n(q) = |q^{-1}(0)|$ and $p(q) = |q^{-1}(1)|$. Then $n(q) + p(q) = |V| = 2^{2k}$.

We claim that

$$n(q + q_0) = 3n(q) + p(q) \text{ and } p(q + q_0) = 3p(q) + n(q). \tag{2.6}$$

Indeed, for $(v, u) \in V \oplus U$, we have $(q + q_0)(v, u) = q(v) + q_0(v)$. As $n(q_0) = 3$ and $p(q_0) = 1$, if $q(v) = 0$, there are three elements $u \in U$ such that $(q + q_0)(v, u) = 0$, and one for which $(q + q_0)(v, u) = 1$. Similarly, if $q(v) = 1$, there are three elements $u \in U$ such that $(q+q_0)(v,u) = 1$, and one for which $(q+q_0)(v,u) = 0$, and the claim follows.

If we set $r(q) = p(q) - n(q)$, then $r(q + q_0) = 2r(q)$ by Equation (2.6). Hence, $r(q)$ and $r(q + q_0)$ have the same sign, proving the inductive step, and concluding the proof of the proposition. □

Theorem 2.33 *For $n \geq 2$, we have $\pi_{n+2}(S^n) \cong \mathbb{Z}_2$.*

Proof We already know that $\pi_4(S^2) \cong \mathbb{Z}_2$ by Remark 2.30, so we can assume that $n \geq 3$. We again use the Pontryagin construction, which gives

$$\pi_{n+2}(S^n) \cong \mathrm{Emb}^{fr}(2,n).$$

Let (M,φ) be a framed surface in $\mathbb{R}^{n+2}$. Given $x \in H_1(M)$, represent it by an embedded, oriented one-manifold $C \subset M$. Let τ be a trivialisation of the normal bundle $\nu_{C \subseteq M}$. Then $(C,(\tau,\varphi))$ is a framed one-manifold and hence represents an element $q(C)$ of

$$\mathrm{Emb}^{fr}(1,n+1) \cong \pi_{n+2}(S^{n+1}) \cong \mathbb{Z}_2,$$

as in the proof of Theorem 2.29, by comparing the framing (τ,φ) to the canonical normal framing of C. Since $n \geq 3$, there is an embedded disk $D \subset \mathbb{R}^{n+2}$ such that $D \cap M = C$. Then $q(C) = 0$ if and only if we can extend the normal framing φ of M to the manifold $M(D)$ obtained by surgering M along D.

If C, $C' \subset M$ are homologous one-manifolds, then $q(C) = q(C')$, so q descends to $H_1(M)$. Pontryagin initially thought q was a homomorphism (which would have implied that $\pi_{n+2}(S^n) = 0$). However, it is a quadratic form associated to the intersection pairing on $H_1(M)$.

Exercise 2.34 Show that

$$q(x+y) = q(x) + q(y) + x \cdot y \tag{2.7}$$

for x, $y \in H_1(M)$, where $x \cdot y$ is the algebraic intersection number of x and y.

In particular, $q(2x) = 2q(x) = 0$, and so q factors through a map

$$\bar{q} \colon H_1(M;\mathbb{Z}_2) \to \mathbb{Z}_2.$$

We set $\mathrm{Arf}(M,\varphi) := \mathrm{Arf}(\bar{q}) \in \mathbb{Z}_2$. We claim that this is a framed cobordism invariant, and hence descends to $\mathrm{Emb}^{fr}(2,n)$. It suffices to prove this for an elementary cobordism. This is true for index zero and three critical points, since then an S^2 component appears or disappears, which does not change H_1. Similarly, H_1 does not change when adding a one-handle between two different components of M. If we add a one-handle $D^1 \times D^2$, then its belt circle

$b = \{0\} \times S^1$ is compressible along $\{0\} \times D^2$, and the framing extends to the core, so $q(b) = 0$. Let a be a curve in M obtained by connecting the endpoints of $D^1 \times \{p\}$ for $p \in \partial D^2$ by an arc a_0 in $M \setminus (\partial D^1 \times D^2)$. We can extend a symplectic basis of $H_1(M;\mathbb{Z}_2)$ disjoint from a_0 with $[a], [b]$, and the Arf invariant remains unchanged as $q(a)q(b) = 0$. Finally, a two-handle addition is the reverse of a one-handle addition, so these also do not change the Arf invariant.

We claim that

$$\mathrm{Arf}\colon \mathrm{Emb}^{fr}(2,n) \to \mathbb{Z}_2$$

is an isomorphism. It is surjective since we can take $F = T^2$, and frame $S^1 \times \{1\}$ and $\{1\} \times S^1$ non-trivially. This extends to the two-cell of T^2 as it is attached along a commutator, and $\pi_1(\mathrm{SO}(n))$ is commutative.

For the injectivity, suppose that $\mathrm{Arf}(M, \varphi) = 0$. Then there is a simple closed curve $C \subsetneq M$ for which $q(C) = 0$ by Proposition 2.32. Since $n \geq 3$, any two embeddings of a surface into $\mathbb{R}^{n+2}$ are isotopic, so we can assume that M is standardly embedded in $\mathbb{R}^3 \subset \mathbb{R}^{n+2}$. In particular, C bounds a disk D in $\mathbb{R}^{n+2}$ along which we can perform surgery to obtain the framed cobordant manifold $M(D)$ of smaller genus, and also having Arf invariant zero. We can keep surgering until we obtain a framed sphere, which is framed null-cobordant as $\pi_2(\mathrm{SO}(n)) = 0$. □

The *suspension* SX of a topological space X is obtained by taking $X \times [-1, 1]$, and collapsing $X \times \{-1\}$ to a point v_- and $X \times \{1\}$ to a point v_+. Given a map $f\colon X \to Y$, it induces a map $Sf\colon SX \to SY$ of suspensions via the formula $(x,t) \mapsto (f(x), t)$ for $x \in X$ and $t \in [-1, 1]$.

Proposition 2.35 *Suspension*

$$s_{k,\ell}\colon \pi_{k+\ell}(S^\ell) \to \pi_{k+\ell+1}(S^{\ell+1})$$

is surjective for $\ell > k$ and is an isomorphism for $\ell > k + 1$.

This follows from the Freudenthal suspension theorem, but we now give an alternative proof using the Pontryagin construction. First, we state a useful result that we will refer to as the 'straightening lemma'.

Lemma 2.36 *Let M be a compact k-manifold embedded in $\mathbb{R}^n$ for $n-k > 1$. If v is a nowhere vanishing normal vector field along M, then there is an isotopy $\{\phi_t : t \in I\}$ of $\mathbb{R}^n$ such that $\phi_0 = Id_{\mathbb{R}^n}$ and $d\phi_1(v)$ is parallel to $\left.\frac{\partial}{\partial x_n}\right|_M$.*

Furthermore, when M has boundary and v is already vertical along ∂M, then the ambient isotopy can be chosen to fix $(\partial M, v|_{\partial M})$ pointwise.

Proof We perturb M such that $x_n|_M$ is Morse. We can assume that v is orthogonal to TM and has unit norm. If $v(p) = -\alpha\frac{\partial}{\partial x_n}$ for $\alpha > 0$, then p is a critical point of $x_n|_M$, which are isolated. We can arrange that v_p never points in the direction of $-\frac{\partial}{\partial x_n}$ as follows. Let $U_p \subset M$ be a small ball around p. Choose a vector field w tangent to the unit normal bundle $U_p \times S^{n-k}$ of U_p that is non-zero at $v(p)$ and is zero outside a small neighbourhood of $v(p)$. Then we rotate v in the direction of w, keeping it unit norm and orthogonal to TM, after which it no longer points in the direction of $-\frac{\partial}{\partial x_n}$ at p.

For $p \in M$, let $\varphi(p) \in (0, \pi]$ be the angle between $v(p)$ and $-\frac{\partial}{\partial x_n}$. Since M is compact, $\varphi_0 := \min(\varphi) > 0$. We rotate $v(p)$ towards $\frac{\partial}{\partial x_n}$ in the plane $\langle v(p), \frac{\partial}{\partial x_n}\rangle$ by angle $\frac{\pi-\varphi(p)}{2}$ at each p, after which we obtain the vector field v'. Then the $\frac{\partial}{\partial x_n}$-component of v' is at least $\varphi_0/2$, and the vector field does not pass through TM during the rotation. (Note that v' is no longer orthogonal to M.) We extend v' to $\mathbb{R}^n$ such that it agrees with $\frac{\partial}{\partial x_n}$ outside an ε-neighbourhood $N(M)$ of M. We let ϕ_t be the flow of v'. Then M leaves $N(M)$ in time $2\varepsilon/\varphi_0$, and outside $N(M)$, the normal vector field v' is of the desired form. □

Proof of Proposition 2.35 We will write $n = k + \ell$. Under the isomorphism of Theorem 2.26, suspension corresponds to mapping a framed k-manifold M in $\mathbb{R}^n \subset S^n$ to $\mathbb{R}^n \times \{0\} \subset \mathbb{R}^{n+1}$ and adding $\left.\frac{\partial}{\partial x_{n+1}}\right|_M$ to the normal framing.

We first show that $s_{k,\ell}$ is surjective when $\ell > k$; that is, $n > 2k$. Let (M, f) be a framed k-manifold in $\mathbb{R}^{n+1}$, where $f = (v_1, \dots, v_{\ell+1})$. Applying Lemma 2.36 to $(M, v_{\ell+1})$, we can assume that $v_{\ell+1}$ is parallel to $\left.\frac{\partial}{\partial x_{n+1}}\right|_M$. Let

$$\pi \colon \mathbb{R}^{n+1} \to \mathbb{R}^n$$

be orthogonal projection along x_{n+1}. Since $n > 2k$, a generic map from a k-manifold into $\mathbb{R}^n$ is an embedding. Hence, by possibly perturbing (M, f) slightly, $\pi(M)$ is also embedded. We can isotope M to $\pi(M)$ linearly in the x_{n+1}-direction. Then $f' := (v_1, \dots, v_\ell)$ is a normal framing of $\pi(M)$, and the suspension of $(\pi(M), f')$ is ambient isotopic, and so framed cobordant, to (M, f).

We now show that $s_{k,\ell}$ is injective when $\ell > k + 1$; that is, $n > 2k + 1$. Let (M, f) be a framed k-manifold in $\mathbb{R}^n$ whose suspension is framed null-cobordant in $\mathbb{R}^{n+1}$ via a framed cobordism (W, f'). We now apply the relative version of the straightening lemma to (W, f') in $I \times \mathbb{R}^{n+1}$, after which we can isotope it into $I \times \mathbb{R}^n$ as in the proof of surjectivity, showing that (M, f) is also framed null-cobordant in $\mathbb{R}^n$. □

Definition 2.37 We write π_k^s for the direct limit $\varinjlim \pi_{k+\ell}(S^\ell)$, and call it the *kth stable homotopy group of spheres*.

Note that Theorems 2.27, 2.29, and 2.33 imply that $\pi_0^s \cong \mathbb{Z}$, $\pi_1^s \cong \mathbb{Z}_2$, and $\pi_2^s \cong \mathbb{Z}_2$, respectively. By the work of Serre [162], the group π_k^s is finite for every integer $k > 0$.

2.8 Surgery Theory and Exotic Spheres

2.8.1 The Group of Homotopy Spheres

The h-cobordism classes of oriented homotopy n-spheres form an Abelian group with respect to connected sum that is usually denoted by Θ_n. The inverse of the homotopy sphere M is given by $-M$. Indeed, $M \# -M$ bounds the manifold $I \times (M \setminus B)$ for an embedded open n-ball $B \subset M$ (where we round the corners along $\partial I \times \partial B$), which is contractible since it contains $M \setminus B$ as a deformation retract.

Exercise 2.38 Show that a simply connected closed n-manifold M is h-cobordant to S^n if and only if it bounds a contractible manifold.

By Proposition 2.8, the manifold $I \times (M \setminus B)$ is actually diffeomorphic to D^{n+1} for $n \geq 5$.

It is clear that $\Theta_1 = \Theta_2 = 0$, and $\Theta_3 = 0$ by the Poincaré conjecture (which is now a theorem due to Perelman). By the h-cobordism theorem, two oriented homotopy spheres of dimension at least five are h-cobordant if and only if they are orientation-preserving diffeomorphic. So, in these dimensions, Θ_n is the set of orientation-preserving diffeomorphism classes of smooth structures on S^n. In dimension four, there are non-trivial h-cobordisms, and, in fact, $\Theta_4 = 0$. However, it is not known whether there are exotic smooth structures on S^4. It is also not clear whether $M \# -M$ is diffeomorphic to S^4 for a homotopy four-sphere M.

Kervaire and Milnor [82] used surgery theory to reduce the computation of the groups Θ_n to understanding homotopy groups of spheres, which allowed them to give explicit formulas in many cases. For example, $\Theta_7 \cong \mathbb{Z}_{28}$. In this section, we will outline the key components of the proof of this result. This section is based on their original paper and the book of Kosinski [89].

2.8.2 Stable Normal and Tangent Bundles

By the Whitney embedding theorem (Theorem 2.2), every smooth n-manifold M embeds into $\mathbb{R}^{2n}$. Consequently, we can embed $I \times M$ into $\mathbb{R}^{2n+2}$, and so any two embeddings of M into $\mathbb{R}^{2n+2}$ are isotopic. In particular, the normal bundle $\nu_{M \subseteq \mathbb{R}^k}$ of M in $\mathbb{R}^k$ for every $k \geq 2n + 2$ is well defined.

Definition 2.39 We say that the vector bundles ξ and η over the same base space are *stably equivalent* if there are $a, b \in \mathbb{N}$ such that $\xi \oplus \varepsilon^a \cong \eta \oplus \varepsilon^b$, where ε^k denotes the rank k trivial bundle.

This is an equivalence relation, and we denote the equivalence class of ξ by $[\xi]$. If we endow the set of equivalence classes with the operation

$$[\xi] + [\eta] := [\xi \oplus \eta],$$

we get a commutative monoid. If we take formal differences $[\xi]-[\eta]$ of equivalence classes (called the *Grothendieck group* of the monoid), we obtain the *reduced K-group* of M, denoted $\widetilde{K}(M)$. This is in fact a ring, where the product is given by $[\xi] \cdot [\eta] := [\xi \otimes \eta]$. Recall that we have encountered the closely related algebraic K-theory in Section 2.4.

For $k < \ell$, we identify $\mathbb{R}^k$ with the subspace $\mathbb{R}^k \times \{\underline{0}\}$ of $\mathbb{R}^\ell$. Then

$$\nu_{M \subseteq \mathbb{R}^\ell} \cong \nu_{M \subseteq \mathbb{R}^k} \oplus \varepsilon^{\ell-k},$$

where the bundle $\varepsilon^{\ell-k}$ is spanned by translates of the standard basis vectors $e_{k+1}, \dots, e_\ell$ along $M \subset \mathbb{R}^k$. So the bundles $\nu_{M \subseteq \mathbb{R}^k}$ for $k \geq 2n+2$ are stably equivalent. We denote by ν_M this equivalence class and call it the *stable normal bundle* of M.

Note that

$$TM \oplus \nu_{M \subseteq \mathbb{R}^k} \cong T\mathbb{R}^k|_M \cong \varepsilon^k.$$

Using the language of K-theory, $\nu_M = -[TM]$.

There is a natural embedding $i_k \colon \mathrm{SO}(k) \to \mathrm{SO}(k+1)$, and we let

$$\mathrm{SO} := \varinjlim SO(k).$$

Proposition 2.40 *The map*

$$i_k^n \colon \pi_n(SO(k)) \to \pi_n(SO(k+1))$$

induced by i_k is an isomorphism for $k > n+1$.

Proof Elements of $\mathrm{SO}(k+1)$ can be viewed as positive orthonormal $(k+1)$-frames $(v_1, \dots, v_{k+1})$. Then $v_{k+1} \in S^k$, and $(v_1, \dots, v_k)$ is a positive orthonormal k-frame in $T_{v_{k+1}} S^k$. Hence, we have a fibration

$$\mathrm{SO}(k) \xrightarrow{i_k} \mathrm{SO}(k+1) \longrightarrow S^k.$$

So, we have the exact sequence

$$\pi_{n+1}(S^k) \longrightarrow \pi_n(\mathrm{SO}(k)) \xrightarrow{i_k^n} \pi_n(\mathrm{SO}(k+1)) \longrightarrow \pi_n(S^k).$$

If $k > n+1$, then $\pi_{n+1}(S^k) = 0$ and $\pi_n(S^k) = 0$, and the result follows. □

Table 2.1 *The stable homotopy groups of SO.*

$n \mod 8$	0	1	2	3	4	5	6	7
$\pi_n(\mathrm{SO})$	$\mathbb{Z}_2$	$\mathbb{Z}_2$	0	$\mathbb{Z}$	0	0	0	$\mathbb{Z}$

By the preceding result, the homotopy groups $\pi_n(\mathrm{SO}(k))$ stabilise as $k \to \infty$ to $\pi_n(\mathrm{SO})$, and $\pi_n(\mathrm{SO}(n+2))$ is already in the stable range; that is,

$$\pi_n(\mathrm{SO}(n+2)) \cong \pi_n(\mathrm{SO}).$$

Remark 2.41 The celebrated *Bott periodicity* theorem gives the homotopy groups of SO as in Table 2.1. Bott proved his result by applying Morse theory to the space of minimal geodesics in SO; see Milnor [114].

Definition 2.42 A manifold M is called *parallelisable* if its tangent bundle TM is trivial. We say that M is *stably parallelisable* if its *stable tangent bundle* $TM \oplus \varepsilon^1$ is trivial.

For example, the sphere S^n is stably parallelisable for every n, since

$$TS^n \oplus \nu_{S^n \subseteq \mathbb{R}^{n+1}} \cong T\mathbb{R}^{n+1}|_{S^n} = \varepsilon^{n+1}$$

and $\nu_{S^n \subseteq \mathbb{R}^{n+1}} \cong \varepsilon^1$.

However, S^n is parallelisable if and only if $n \in \{0, 1, 3, 7\}$ by the work of Adams [1]. To see that these spheres are parallelisable, note they are the unit spheres of $\mathbb{R}$, the complex plane $\mathbb{C} \cong \mathbb{R}^2$, the quaternions $\mathbb{H} \cong \mathbb{R}^4$, and the Cayley octonions $\mathbb{O} \cong \mathbb{R}^8$, respectively, which are the only division algebras.

Theorem 2.43 *Every homotopy sphere is stably parallelisable.*

Proof We only give the proof in dimension n when $n \geq 6$ and $n \equiv 3, 5, 6$, or 7 mod 8. A homotopy n-sphere is homeomorphic to S^n for $n \geq 6$ by the Generalised Poincaré Conjecture (Theorem 2.9). A vector bundle of rank $n+1$ over S^n is determined by its clutching function $S^{n-1} \to \mathrm{SO}(n+1)$. However, $\pi_{n-1}(\mathrm{SO}(n+1)) = 0$ when $n \equiv 3, 5, 6$, or 7 mod 8 by Bott periodicity (see Remark 2.41), so the vector bundle is trivial. □

Lemma 2.44 *Let ξ be a rank k vector bundle over an n-complex B, such that $k > n$. If $\xi \oplus \varepsilon^r \cong \varepsilon^{k+r}$ for some r, then $\xi \cong \varepsilon^k$.*

Proof We may assume that $r = 1$. Indeed, if we apply the $r = 1$ case to the isomorphism

$$(\xi \oplus \varepsilon^{r-1}) \oplus \varepsilon^1 \cong \varepsilon^{k+r},$$

we obtain that $\xi \oplus \varepsilon^{r-1} \cong \varepsilon^{k+r-1}$. Repeating this process, we see that $\xi \cong \varepsilon^k$.

The isomorphism $\xi \oplus \varepsilon^1 \cong \varepsilon^{k+1}$ implies that there is a map $f\colon B \to S^k$ such that $\xi \cong f^*TS^k$. Since $\dim(B) = n < k$, the map f is null-homotopic, and hence ξ is trivial. □

Corollary 2.45 *Let M be an n-dimensional submanifold of S^{n+k} for $n < k$. Then M is stably parallelisable if and only if its normal bundle ν is trivial.*

Proof Since $TM \oplus \nu$ is trivial, so is $(TM \oplus \varepsilon^1) \oplus \nu$. The claim now follows from Lemma 2.44, applied with the roles $\xi = TM \oplus \varepsilon^1$ and $\xi = \nu$ for the two directions. □

Corollary 2.46 *Let W be a connected n-manifold with $\partial W \neq \emptyset$. Then W is stably parallelisable if and only if it is parallelisable.*

Proof Since $\partial W \neq \emptyset$, it is homotopy equivalent to an $(n-1)$-complex. As TW is n-dimensional, the result follows from Lemma 2.44. □

Definition 2.47 We say that the closed, connected manifold M is *almost parallelisable* if $T(M \setminus \{p\})$ is trivial for some point $p \in M$.

For example, the sphere S^n is almost parallelisable for every n as $S^n \setminus \{p\} \approx \mathbb{R}^n$ is parallelisable.

Exercise 2.48 Show that every closed, orientable surface is almost parallelisable.

By Corollary 2.46, every stably parallelisable manifold is almost parallelisable. Conversely, let M be an almost parallelisable n-manifold. If $4 \nmid n$, then M is stably parallelisable. If $4 \mid n$, then M is stably parallelisable if and only if $\sigma(M) = 0$.

2.8.3 The J-Homomorphism

The *J-homomorphism*

$$J_{k,\ell}\colon \pi_k(\mathrm{SO}(\ell)) \to \pi_{k+\ell}(S^\ell)$$

was defined in full generality by Whitehead [177], building on the work of Hopf. Recall that we can write

$$S^{k+\ell} \approx \partial(D^{k+1} \times D^\ell) \approx (S^k \times D^\ell) \cup (D^{k+1} \times S^{\ell-1}).$$

Hence, we can view $S^k \times \{0\}$ as a sphere embedded in $S^{k+\ell}$ with normal framing $S^k \times D^\ell$. Given a map $f\colon S^k \to \mathrm{SO}(\ell)$, we can twist the normal framing using f. The resulting framed sphere gives rise to an element $[j_{k,\ell}(f)] \in \pi_{k+\ell}(S^\ell)$ via the Pontryagin construction; see Section 2.7. Explicitly, if $(x, y) \in S^k \times D^\ell \subseteq S^{k+\ell}$, we let

$$j_{k,\ell}(f)(x, y) = [f(x)(y)] \in S^\ell \approx D^\ell/S^{\ell-1},$$

while $j_{k,\ell}(f)(x, y)$ is the image of $S^{\ell-1}$ in $D^\ell/S^{\ell-1}$ for $(x, y) \in D^{k+1} \times S^{\ell-1} \subseteq S^{k+\ell}$. The homotopy class of $j_{k,\ell}(f)$ only depends on the homotopy class of f; we set $J_{k,\ell}([f]) := [j_{k,\ell}(f)]$.

Recall that the groups $\pi_k(\mathrm{SO}(\ell))$ and $\pi_{k+\ell}(S^\ell)$ stabilise as $\ell \to \infty$; see Propositions 2.35 and 2.40. The maps $J_{k,\ell}$ stabilise once $\ell > k + 1$, and we write $J_k := J_{k,\ell}$ for ℓ in the stable range. It is a map

$$J_k\colon \pi_k(\mathrm{SO}) \to \pi_k^s$$

that we also call J-homomorphism. By Bott periodicity (Remark 2.41), the group $\pi_k(SO)$ is cyclic, and by Serre [162], the group π_k^s is finite for $k > 0$. Let

$$j_k := |\mathrm{Im}(J_k)|.$$

We write

$$t_k := 2^{2k-1}(2^{2k-1} - 1)B_k j_{4k-1} a_k/k,$$

where

$$a_k := \begin{cases} 1 & \text{if } k \text{ is even,} \\ 2 & \text{if } k \text{ is odd,} \end{cases}$$

and B_k is the kth Bernoulli number defined in the discussion preceding Theorem 1.67. For example, $B_2 = 1/30$. Furthermore, $j_7 = 240$ by the work of Adams [2]. Hence

$$t_2 = 8 \cdot 7 \cdot \frac{1}{30} \cdot 240 \cdot \frac{1}{2} = 8 \cdot 28, \tag{2.8}$$

which we will use in the computation of Θ_7.

Theorem 2.49 *The signatures of $4k$-dimensional almost parallelisable manifolds form the group $t_k\mathbb{Z}$.*

Proof If M and M' are almost parallelisable, then so is $M \# M'$, and we have $\sigma(M \# M') = \sigma(M) + \sigma(M')$. Hence signatures of almost parallelisable manifolds form a group.

Let M be an almost parallelisable $4k$-manifold, and write $n = 4k$. We can embed M into $\mathbb{R}^{2n+2}$ such that $M \cap \mathbb{R}^{2n+2}_-$ is diffeomorphic to D^n and $M \cap \mathbb{R}^{2n+1} = S^{n-1}$. Then $N := M \cap \mathbb{R}^{2n+2}_+$ is parallelisable, so its normal bundle ν_N is trivial by Corollary 2.45. Let φ_N be a trivialisation of ν_N. Comparing the trivialisation $\varphi_S := \varphi_N|_{S^{n-1}}$ to the standard framing of $\nu_{S^{n-1} \subset \mathbb{R}^{2n+1}}$, we obtain a map $f \colon S^{n-1} \to \mathrm{SO}(n+2)$ such that

$$[f] \in \operatorname{Ker}(J_{n-1,n+2}) \cong \operatorname{Ker}(J_{n-1}) \leq \pi_{n-1}(\mathrm{SO}),$$

since (S^{n-1}, φ_S) is framed null-cobordant via (N, φ_N). Furthermore,

$$\pi_{n-1}(\mathrm{SO}) \cong \mathbb{Z}$$

by Bott periodicity (see Remark 2.41), hence $\operatorname{Ker}(J_{n-1}) = j_{n-1}\mathbb{Z}$.

Fix a diffeomorphism $\varphi_D \colon M \cap \mathbb{R}^{2n+2}_- \to D^n$ that is the identity on S^{n-1}. We can view $p := \varphi_D^{-1}(\{0\}) \in M$ as a framed point in M. If we apply the Pontryagin construction to $(\{p\}, \varphi_D)$, we obtain a degree one map $g \colon M \to S^n$; that is, $g_*([M]) = [S^n]$. Let ξ be the rank $n+2$ bundle over S^n with clutching function f. Then $g^*\xi$ is the normal bundle ν_M of M. Hence, by the naturality of Pontryagin classes,

$$p_k(\nu_M)[M] = \langle g^* p_k(\xi), [M] \rangle = \langle p_k(\xi), g_*([M]) \rangle = p_k(\xi)[S^n]. \tag{2.9}$$

We now use a result of Kervaire [79, lemma 1.1] that states that

$$p_k(\xi)[S^n] = a_k(2k-1)![f] \tag{2.10}$$

if ξ is a bundle of rank greater than n over S^n with clutching function $[f] \in \pi_{n-1}(\mathrm{SO}) \cong \mathbb{Z}$.

As TM is trivial over the $(n-1)$-skeleton of M, all Pontryagin classes $p_\ell(M) = 0$ for $\ell < k$, and the total Pontryagin class is $p(M) = 1 + p_k(M)$. It follows from Corollary 2.45 that the normal bundle ν_M is also trivial over the $(n-1)$-skeleton of M, hence similarly $p(\nu_M) = 1 + p_k(\nu_M)$. Since $TM \oplus \nu_M = \varepsilon^{2n+2}$, we have

$$2p(M)p(\nu_M) = 2p(\varepsilon^{2n+2}) = 0;$$

see Equation (A.1) in Section A.3.4. On the other hand,

$$p(M)p(\nu_M) = 1 + p_k(M) + p_k(\nu_M).$$

Since $H^{4k}(M)$ has no two-torsion, we conclude that

$$p_k(M)[M] = -p_k(\nu_M)[M] = -a_k(2k-1)![f],$$

where the second equality follows from Equations (2.9) and (2.10). Hence, by the Hirzebruch signature theorem (Theorem 1.67),

$$\begin{aligned}\sigma(M) =& p_k(M)[M]2^{2k}(2^{2k-1}-1)B_k/(2k)! = \\ &-2^{2k-1}(2^{2k-1}-1)B_k a_k[f]/k.\end{aligned} \tag{2.11}$$

As $[f]$ is divisible by j_{n-1}, we conclude that $\sigma(M)$ is divisible by t_k.

To obtain an almost parallelisable n-manifold M with $\sigma(M) = t_k$, frame S^{n-1} with $[f]$ generating $\mathrm{Ker}(J_{n-1})$. This framing extends to a manifold $M_1 \subset \mathbb{R}^{2n+2}$. If we set $M := M_1 \cup_\partial D^n$, then either $\sigma(M)$ or $\sigma(-M) = -\sigma(M)$ agrees with t_k by Equation (2.11). □

2.8.4 The Surgery Exact Sequence

In Section 2.8.1, we introduced the group Θ_n of h-cobordism classes of homotopy n-spheres with the connected sum operation. By Theorem 2.43, every homotopy sphere is stably parallelisable, so has a trivial stable normal bundle. If we choose a normal framing, we obtain an element of $\mathrm{Emb}^{fr}(n,k)$ for $k > n+1$. By the Pontryagin construction (Theorem 2.26), we have

$$\mathrm{Emb}^{fr}(n,k) \cong \pi_{n+k}(S^k) \cong \pi_n^s.$$

The choice of normal framing is not unique, but any two framings can be shown to differ by taking the connected sum with a framed S^n. Hence, the image of a homotopy n-sphere gives a well-defined element of

$$\mathrm{coker}(J_n) = \pi_n^s/\mathrm{Im}(J_n).$$

In fact, we obtain a homomorphism

$$h\colon \Theta_n \to \mathrm{coker}(J_n).$$

Let bP_{n+1} be the subgroup of Θ_n consisting of homotopy spheres that bound parallelisable manifolds.

Lemma 2.50 *We have $\mathrm{Ker}(h) = bP_{n+1}$.*

Proof First, suppose that $M \in \mathrm{bP}_{n+1}$. Then there is a parallelisable $(n+1)$-manifold W such that $\partial W = M$. As W is parallelisable, its stable normal bundle is trivial by Corollary 2.45; let f be a trivialisation. Then (W, f) is a framed null-cobordism of $(M, f|_M) \in \mathrm{Emb}^{fr}(n,k)$, hence

$$h([M]) = 0 \in \mathrm{coker}(J_n),$$

where $[M]$ is the h-cobordism class of M.

Conversely, suppose that $h([M]) = 0$. Then there is a stable normal framing f_0 of M such that (M, f_0) is framed null-cobordant. In other words, there is an $(n+1)$-manifold W with a stable normal framing f such that $f|_M = f_0$. Then W is parallelisable by Corollaries 2.45 and 2.46. □

We denote the image of h by Σ_n^s. Using the Pontryagin construction, this is the subgroup of $\mathrm{Emb}^{fr}(n,k) \cong \pi_n^s$ consisting of elements representable by homotopy spheres. Hence, we have obtained the following result:

Theorem 2.51 *The following sequence is exact:*

$$0 \longrightarrow bP_{n+1} \longrightarrow \Theta_n \xrightarrow{h} coker(J_n) \longrightarrow \pi_n^s/\Sigma_n^s. \tag{2.12}$$

The term $\mathrm{coker}(J_n)$ is purely homotopy-theoretic. For example, we have $\mathrm{coker}(J_7) = 0$, so $\Theta_7 \cong \mathrm{bP}_8$.

What remains is to determine the groups bP_n and π_n^s/Σ_n^s. This is where surgery comes into the picture.

Theorem 2.52 *Let $n > 4$, and suppose that $n \notin \{6, 14\}$. Then $\pi_n^s/\Sigma_n^s = 0$ if $n \not\equiv 2 \mod 4$, and $|\pi_n^s/\Sigma_n^s| \le 2$ otherwise.*

The result in the case $n \equiv 2 \mod 4$ is shown by constructing a homomorphism $\kappa\colon \pi_n^s \to \mathbb{Z}_2$ with kernel Σ_n^s, called the *Kervaire invariant*. This is the obstruction to eliminating the homotopy groups of a manifold representing a class in $\pi_n^s \cong \mathrm{Emb}^{fr}(n,k)$ using a sequence of framed surgeries; that is, a framed cobordism to a homotopy sphere. We have already encountered it for $n = 2$ in the proof of Theorem 2.33. When M is a $(4k+2)$-manifold with a normal framing φ, it is defined as the Arf invariant (Definition 2.31) of a quadratic refinement of the intersection form on $H_{2k+1}(M;\mathbb{Z}_2)$ induced by φ. Note that the intersection form on an odd-dimensional homology group is antisymmetric, and κ can be viewed as an odd-dimensional analogue of the signature. We now define κ when $k \notin \{0,1,3\}$, in which case it does not depend on the normal framing.

Definition 2.53 Let M be a $(4k+2)$-dimensional almost parallelisable manifold, and suppose that $k \notin \{0,1,3\}$. By Theorem 2.60, we can kill all the homotopy groups of M below dimension $2k+1$ using surgery. We can represent every $x \in H_{2k+1}(M)$ by an embedded sphere S_x by a result of Haefliger. We define $q(x) = 0$ if $\nu_{S \subset M}$ is trivial, and $q(x) = 1$ otherwise. Then q is a quadratic refinement of the intersection form on $H_{2k+1}(M)$. We define the Kervaire invariant $\kappa(M)$ as $\mathrm{Arf}(q)$.

Browder [13] proved that manifolds with Kervaire invariant one can only exist in dimensions $2^k - 2$. There are examples in dimensions 2, 6, 14, 30, and 62, and Hill, Hopkins, and Ravenel [63] further showed that the only possible dimensions are these, and maybe 126, which is still open.

Theorem 2.54 *The group bP_{n+1} is finite cyclic, and*

$$|bP_{n+1}| = \begin{cases} t_k/8 & \text{if } n = 4k - 1 \text{ and } k > 1, \\ 1 \text{ or } 2 & \text{if } n = 4k + 1, \\ 1 & \text{if } n \in \{1, 5, 13\}, \\ 1 & \text{if } n = 2k. \end{cases}$$

The idea of the proof is again to start with a parallelisable $(n + 1)$-manifold bounding a homotopy sphere M and try to kill all its homotopy groups via surgery. If this succeeds, we obtain a homotopy ball bounding M, showing that M is diffeomorphic to S^n by the h-cobordism theorem (Proposition 2.8). We develop the necessary techniques and prove the case $n = 4k - 1$ in the following section.

2.8.5 Killing Homotopy Groups Using Surgery

In this section, we summarise the effect of surgery (Definition 1.58) along a framed sphere on the homology of a manifold. Our goal is to kill the homology using a sequence of surgeries. This is easy below the middle dimension, but we might run into obstructions in the middle dimension. Once the homology vanishes up to the middle dimension, it also vanishes above by Poincaré duality.

Proposition 2.55 *Let M be an n-manifold, and $S \subset M$ an embedded k-sphere with trivial normal bundle and normal framing ν. Then*

$$H_i(M(S, \nu)) \cong \begin{cases} H_i(M) & \text{if } i < k \text{ and } n \geq 2k + 1, \\ H_i(M)/[S] & \text{if } i = k \text{ and } n \geq 2k + 2. \end{cases}$$

The following result describes how the fundamental group changes under surgery along a one-sphere.

Lemma 2.56 *Let M be an n-manifold, and $S \subset M$ an embedded 1-sphere with trivial normal bundle and normal framing ν. If $n > 3$, then*

$$\pi_1(M(S,\nu)) \cong \pi_1(M)/\langle [S] \rangle,$$

where $\langle [S] \rangle$ is the normal subgroup generated by the homotopy class of S.

We now describe the effect of surgery in the 'middle dimension'.

Proposition 2.57 *If $\dim(M) \in \{2k, 2k+1\}$ and S is a framed k-sphere representing a primitive homology class, then surgery along S kills $[S]$ and does not change $H_i(M)$ for $i < k$.*

The following key lemma ensures that every homotopy class below the middle dimension can be represented by a sphere with trivial normal bundle.

Lemma 2.58 *If M is a stably parallelisable n-manifold and $k < n/2$, then every $\alpha \in \pi_k(M)$ can be represented by an embedded sphere with trivial normal bundle.*

Proof Since $n \geq 2k+1$, the space of embeddings of S^k into M is dense in the space of smooth maps, and so there is an embedded sphere S in M representing α; see Section 1.3.

As M and S are stably parallelisable,

$$\nu_{S \subset M} \oplus \varepsilon^{k+1} \cong \nu_{S \subset M} \oplus TS \oplus \varepsilon^1 \cong TM|_S \oplus \varepsilon^1 \cong \varepsilon^{n+1}.$$

As the rank of $\nu_{S \subset M}$ is $n-k > k = \dim(S)$, it follows from Lemma 2.44 that $\nu_{S \subset M} \cong \varepsilon^{n-k}$, as claimed. □

The following result gives sufficient conditions for a framing of the stable tangent bundle to extend to the manifold obtained after the surgery:

Proposition 2.59 *Let f be a framing of the stable tangent bundle of an n-manifold M, and let S be a k-sphere in M with trivial normal bundle. If either $k < n/2$ or $k = n/2$ and $n \notin \{2, 6, 14\}$, then there is a normal framing ν of S such that f extends to $M(S,\nu)$.*

Proof Let $W := W(S,\nu)$ be the trace of the surgery on M along the framed sphere (S,ν), which is a cobordism from M to $M(S,\nu)$. The framing f of the stable tangent bundle

$$TM \oplus \varepsilon^1 \cong TW|_M$$

extends to $I \times M$. The obstruction to extending this further to the handle

$$h \approx D^{k+1} \times D^{n-k}$$

is obtained by comparing f along $\{1\} \times S$ to the homotopically unique framing of Th, giving rise to an element $o(f, v) \in \pi_k(\mathrm{SO}(n+1))$. If this element vanishes, f extends to the core $D^{k+1} \times \{0\}$, from which we can further extend it to h.

Note that $\pi_k(\mathrm{SO}(n-k))$ acts freely and transitively on normal framings of S. If we change v by $a \in \pi_k(\mathrm{SO}(n-k))$, then

$$o(f, a \cdot v) = i_*(a) \cdot o(f, v),$$

where $i_* : \pi_k(\mathrm{SO}(n-k)) \longrightarrow \pi_k(\mathrm{SO}(n+1))$ is the homomorphism induced by the usual embedding $i : \mathrm{SO}(n-k) \hookrightarrow \mathrm{SO}(n+1)$. Indeed,

$$Th|_{D^{k+1} \times \{0\}} \cong TD^{k+1} \oplus \nu_{D^{k+1} \subset h}.$$

Under the assumptions of the proposition, the map i_* is surjective (for $k < n/2$, this follows from the proof of Proposition 2.40), and so we can find a framing v for which $o(f, v) = 0$. □

Theorem 2.60 *Let M be a compact, stably parallelisable manifold of dimension $n \geq 2k$. By a sequence of surgeries on M, one can obtain a stably parallelisable manifold M_1 which is $(k-1)$-connected.*

Proof We proceed by induction on k. If $k = 1$, we can obtain a connected manifold by taking the connected sum of the components of M. A connected sum is the same as surgery along the zero-sphere whose points are the connected sum points.

For the inductive step, suppose that M is $(i-1)$-connected for $i < k$. If $i > 1$, by the Hurewicz theorem, the homomorphism

$$\pi_i(M) \longrightarrow H_i(M)$$

sending the homotopy class of a based map $\varphi : S^i \longrightarrow M$ to $\varphi_*([S^i]) \in H_i(M)$ is an isomorphism. Hence, every $\alpha \in H_i(M)$ can be represented by an embedded sphere S with trivial normal bundle by Lemma 2.58. Surgery along S preserves $H_j(M)$ for $j < i$ and kills $[S] \in H_i(M)$ according to Lemma 2.55. The resulting manifold is also stably parallelisable for a suitable normal framing of S by Proposition 2.59. We then repeat this procedure until $H_i(M) = 0$, and hence $\pi_i(M) = 0$ by the Hurewicz theorem.

The case $i = 1$ proceeds analogously, with the only difference that we directly kill $\pi_1(M)$, which is finitely generated, using Lemma 2.56. □

Hence, we can always kill the homotopy groups of a manifold up to the middle dimension using surgery. However, it is not always possible to kill the middle homotopy groups. There are two possible issues. Firstly, a homotopy class might not be representable by an embedded sphere with trivial normal bundle. By Lemma 2.58, this only arises when n is even and the homotopy class is $n/2$-dimensional. Secondly, it is possible that the surgery does not simplify the middle-dimensional homology group. By Proposition 2.57, simplification requires the sphere to represent a primitive homology class. So we can only reduce to the torsion subgroup of the middle-dimensional homology, which can be non-trivial for odd-dimensional manifolds.

For a $4m$-manifold M that is either closed or has a homotopy sphere boundary, the intersection form is symmetric and non-degenerate, and so we can define its signature $\sigma(M)$.

Theorem 2.61 *Let M be a stably parallelisable $4m$-manifold for $m > 1$, such that ∂M is either empty or a homotopy sphere. Then there is a sequence of surgeries from M to a homotopy sphere or disk if and only if $\sigma(M) = 0$.*

Proof It suffices to show that we can obtain a $2m$-connected manifold M_1 using surgeries, since such an M_1 is a homotopy sphere or disk by Poincaré duality and the Hurewicz theorem.

As the signature is unchanged by surgery (i.e., it is a cobordism invariant), the condition $\sigma(M) = 0$ is necessary. The fact that M has boundary can be dealt with by attaching a cone over ∂M, obtaining a closed homology manifold with the same signature.

In the opposite direction, suppose that $\sigma(M) = 0$. By Theorem 2.60, we can assume that M is $(2m-1)$-connected.

Since TM is stably parallelisable, $\pi_1(M) = 1$, and $\dim(M) = 4m \geq 6$, a generalisation of the Whitney embedding theorem (Theorem 2.2) due to Haefliger shows that every $x \in H_{2m}(M) \cong \pi_{2m}(M)$ has an embedded sphere S representing it. This relies on first finding a bundle monomorphism $TS^{2m} \to TM$ covering a continuous map $S^{2m} \to M$ in the homotopy class x, integrating this to an immersion using Haefliger's result, and finally eliminating the double points using the Whitney trick.

The intersection form on $H_{2m}(M)$ is even; that is, $x \cdot x$ is even for every $x \in H_{2m}(M)$. Indeed, an embedded sphere S representing x has trivial stable normal bundle, and hence vanishing Stiefel–Whitney classes. So

$$x \cdot x = \langle e(\nu_{S \subset M}), [S] \rangle \equiv \langle w_{2m}(\nu_{S \subset M}), [S] \rangle \equiv 0 \mod 2.$$

Furthermore, the intersection form on $H_{2m}(M)$ has determinant ± 1 by Poincaré duality, and signature zero by assumption. Hence, $H_{2m}(M)$ has a symplectic basis

$$\lambda_1, \ldots, \lambda_r, \mu_1, \ldots, \mu_r;$$

that is, $\lambda_i \cdot \lambda_j = 0 = \mu_i \cdot \mu_j$ and $\lambda_i \cdot \mu_j = \delta_{ij}$.

Let S be an embedded $2m$-sphere in M representing λ_1. The normal bundle $\nu_{S \subset M}$ of S is stably trivial; we claim that it is actually trivial. Choose a trivialisation $\nu_{S \subset M} \oplus \varepsilon^1 \cong S^{2m} \times \mathbb{R}^{2m+1}$. The ε^1 summand is spanned by a section of $S^{2m} \times \mathbb{R}^{2m+1}$ consisting of unit vectors; that is, a map $u\colon S^{2m} \to S^{2m}$. Then the orthogonal sub-bundle $u^\perp$ is isomorphic to $\nu_{S \subset M}$. As $[S] \cdot [S] = 0$, the normal Euler class $e(\nu_{S \subset M})$ vanishes. Since

$$\langle e(TS^{2m}), [S^{2m}] \rangle = \chi(S^{2m}) = 2,$$

we have $e(TS^{2m}) = 2\mathrm{PD}([S^{2m}])$. Using $u^\perp = u^* TS^{2m}$ and the naturality of the Euler class,

$$0 = e(\nu_{S \subset M}) = e(u^* TS^{2m}) = u^*(e(TS^{2m})) = u^*(2\mathrm{PD}([S^{2m}])).$$

But $u^*(\mathrm{PD}([S^{2m}])) = 0$ is the degree of the map u, which is hence null-homotopic, and so $\nu_{S \subset M} \cong u^* TS^{2m}$ is trivial, as claimed.

The manifold $M(S, \nu)$ is also stably parallelisable by Proposition 2.59 for a suitable normal framing ν of S. As $\lambda_1 \cdot \mu_1 = 1$, the class $[S]$ is primitive. According to Proposition 2.57, the surgered manifold $M(S, \nu)$ is $(2m-1)$-connected, and

$$\mathrm{rk}\,(H_{2m}(M(S, \nu))) < \mathrm{rk}\,(H_{2m}(M)).$$

Thus, we can kill H_{2m} using finitely many surgeries. □

Corollary 2.62 *If $m > 1$ and a $(4m-1)$-dimensional homotopy sphere S bounds a stably parallelisable manifold M with $\sigma(M) = 0$, then S is diffeomorphic to S^{4m-1}.*

Proof First, S bounds a contractible manifold M_1 by Theorem 2.61. Then M_1 is diffeomorphic to D^{4m} by Proposition 2.8, and hence $\partial M_1 = S$ is diffeomorphic to S^{4m-1}. □

We are now ready to prove the $4m$-dimensional case of Theorem 2.54.

Proposition 2.63 *The group bP_{4m} is finite cyclic of order $t_m/8$ for $m > 1$.*

Proof Let P_n be the commutative monoid of compact, oriented, framed n-manifolds with homotopy sphere boundaries, up to framed cobordism that is trivial along the boundary. (This equivalence relation is generated by framed surgery along the interior.) The operation is boundary connected sum. If $n > 5$, then P_n is a group, since $(-M,\varphi)$ is the inverse of (M,φ) by the h-cobordism theorem. Let P_n^0 be the subgroup of P_n consisting of manifolds with boundary S^{n-1}; that is, the kernel of the boundary map $P_n \to bP_n$. Then we have a short exact sequence

$$0 \to P_n^0 \to P_n \to bP_n \to 0. \tag{2.13}$$

Signature $\sigma\colon P_{4m} \to \mathbb{Z}$ is a homomorphism that is injective, since $\sigma(M) = 0$ implies that M is framed cobordant to a contractible manifold by Theorem 2.61, and hence M is diffeomorphic to D^{4m} by Proposition 2.8. We claim that

$$\sigma(P_{4m}) = 8\mathbb{Z}.$$

It is an algebraic result that the signature of any unimodular, even, symmetric matrix over $\mathbb{Z}$ is divisible by eight. To obtain a manifold with signature eight, we introduce the notion of plumbing.

Definition 2.64 Let $p_i\colon E_i \to M_i$ be a bundle with fibre D^k and base a k-manifold M_i for $i \in \{1,2\}$. Furthermore, let $D_i \subset M_i$ be an embedded k-disk, together with a trivialisation $h_i\colon D^k \times D^k \to p_i^{-1}(D_i)$. Then the *plumbing* of E_1 and E_2 is obtained from $E_1 \sqcup E_2$ by identifying $h_1(x,y)$ and $h_2(y,x)$ for every $(x,y) \in D^k \times D^k$.

Given a tree T with vertices labelled by integers, called a *plumbing graph*, one can form a $2k$-manifold by taking a D^k-bundle over S^k for each vertex whose Euler class is given by the label of the vertex, and plumbing bundles together if the corresponding vertices are connected by an edge.

The intersection form of a $2k$-manifold obtained from a plumbing graph with vertices $v_1,\dots,v_r$ is given by an $r \times r$ matrix whose (i,j)-th entry for $i \neq j$ is 1 if $(i,j) \in E(T)$ and is 0 otherwise, and whose (i,i)-th entry is the label of v_i.

The E_8 *graph* consists of a line of seven vertices plus an additional leaf connected to the fifth vertex, all labelled by two. Let $M(4m)$ be the plumbing of the unit disk bundles DTS^{2m} along the E_8 graph. Note that $\chi(S^{2m}) = 2$, so the Euler class of DTS^{2m} is $2\mathrm{PD}([S^{2m}])$, as required. Then the intersection form of $M(4m)$ is given by Equation (6.1), and hence $\sigma(M(4m)) = 8$. So σ is indeed surjective onto $8\mathbb{Z}$.

By Theorem 2.49, signatures of almost parallelisable closed $4m$-manifolds form the group $t_m\mathbb{Z}$. This agrees with $\sigma(P_{4m}^0)$, since gluing D^{4m} gives a bijection between parallelisable manifolds bounding S^{4m-1} and closed, almost parallelisable manifolds.

In conclusion, $\sigma\colon P_{4m} \to 8\mathbb{Z}$ is an isomorphism and $\sigma(P^0_{4m}) = t_m\mathbb{Z}$, so it follows from the exact sequence (2.13) that $bP_{4m} \cong \mathbb{Z}_{t_m/8}$. □

Theorem 2.65 *We have* $\Theta_7 \cong \mathbb{Z}_{28}$.

Proof It follows from the surgery exact sequence (2.12) and the fact that $\operatorname{coker}(J_7) = 0$ that $\Theta_7 \cong \mathrm{bP}_8$. Furthermore, $\mathrm{bP}_8 \cong \mathbb{Z}_{t_2/8}$ by Proposition 2.63, and $t_2/8 = 28$ by Equation (2.8). □

Definition 2.66 Let $a_1, \ldots, a_n \geq 2$ be integers. Then we define the *Brieskorn manifold*

$$M(a_1, \ldots, a_n) := \{ (z_1, \ldots, z_n) \in \mathbb{C}^n\colon z_1^{a_1} + \cdots + z_n^{a_n} = 0 \} \cap S^{2n-1}.$$

This is the boundary of the manifold

$$V(a_1, \ldots, a_n) := \{ (z_1, \ldots, z_n) \in \mathbb{C}^n\colon z_1^{a_1} + \cdots + z_n^{a_n} = \varepsilon \} \cap D^{2n}$$

for $\varepsilon > 0$ sufficiently small.

Exercise 2.67 Show that $M(a_1, \ldots, a_n)$ is a manifold of dimension $2n - 3$.

Brieskorn [12] and Pham [145] proved the following:

Theorem 2.68 *The manifold* $V(a_1, \ldots, a_n)$ *is* $(n-1)$*-connected and parallelisable, and* $M(a_1, \ldots, a_n)$ *is* $(n-2)$*-connected.*

For $j \geq 1$*, the Brieskorn manifolds*

$$\Sigma_j := M(6j - 1, 3, 2, 2, 2) \subset S^9$$

give representatives of $\Theta_7 \cong \mathbb{Z}_{28}$ *depending on* $j \mod 28$.

More generally,

$$M(6j - 1, 3, 2, \ldots, 2) \subset S^{4m+1}$$

is diffeomorphic to j *times the generator of* bP_{4m}.

Definition 2.69 Let M and N be smooth manifolds and $f, g\colon M \looparrowright N$ immersions. Then f and g are *regularly homotopic* if they lie in the same path component of the space of immersions of M into N, endowed with the C^∞ topology.

Exercise 2.70 Given an immersion $f\colon S^1 \looparrowright \mathbb{R}^2$, its *winding number* is the degree of the map $f'/|f'|\colon S^1 \to S^1$. Show that two immersions from S^1 to

$\mathbb{R}^2$ are regularly homotopic if and only if they have the same winding number. This is known as the Whitney–Graustein theorem.

Ekholm and Szűcs [30] showed that, for each congruence class $a \in \mathbb{Z}_{28}$, the embeddings $\Sigma_j \subset S^9$ for $j \equiv a \mod 28$ are pairwise not regularly homotopic, and represent all regular homotopy classes of the homotopy sphere Σ_j that contain an embedding.

2.9 Cerf Theory

This section is based on the papers of Cerf [21] and Kirby [83], and notes of Etnyre [32]. Recall that the proof of the h-cobordism theorem relied on choosing a Morse function on the cobordism and cancelling all of its critical points. Let

$$\{f_t : t \in I\}$$

be a generic one-parameter family of smooth functions on an n-manifold M. Then there is a finite set $B \subset I$ such that, for every $t \in I \setminus B$, the function f_t is Morse with distinct critical values (i.e., $f_t|_{\mathrm{Crit}(f_t)}$ is injective). For $b \in B$, there are two possibilities. The function f_b either has two critical points with the same value, and, as t passes b, the corresponding pair of critical values cross each other. Or f_b has a single degenerate critical point p of the form

$$f_b(x_1, \dots, x_n) = f(p) - x_1^2 - \cdots - x_i^2 + x_{i+1}^2 + \cdots + x_{n-1}^2 + x_n^3$$

in a neighbourhood of p. We call p a *birth/death critical point* and i the *index* of p. There is an $\varepsilon > 0$ and local coordinates $(x_1, \dots, x_n)$ about p such that

$$f_t(x_1, \dots, x_n) = -x_1^2 - \cdots - x_i^2 + x_{i+1}^2 + \cdots + x_{n-1}^2 + x_n^3 \pm tx_n$$

for $t \in (b - \varepsilon, b + \varepsilon)$. Hence, as t passes b, a pair of index i and $i + 1$ critical points appear when the coefficient of the term tx_n is -1, or disappear otherwise. Often, this result in singularity theory is called ‘Cerf theory’. However, this is a misnomer, as it is due to Thom [169], Mather [105][106][107][108], and Smale, predating the work of Cerf.

Cerf theory [21] is an extension to one-parameter families of smooth functions of the methods we used in the proof of the h-cobordism theorem to manipulate the critical points of a single Morse function. More concretely, suppose that all the f_t take values in a compact interval J. Cerf gave a set of moves to manipulate the so-called *Cerf graphic* of the family, defined as the subset $S \subset I \times J$ consisting of pairs (t, u) such that u is a critical value of f_t. Note that S has cusp singularities at those (t, u) where u is the value of a birth/death critical point of f_t, and is otherwise an immersed one-manifold with at most transverse double points and non-vertical tangents. Furthermore, we label each component of S minus the double points and cusps by the index of the corresponding critical point.

For example, when f_0 and f_1 have no singularities, $\pi_1(M) = 1$, and

$$\dim(M) \geq 6,$$

Cerf showed that one can eliminate S completely using his moves without altering f_0 and f_1. This has the following important consequence:

Theorem 2.71 *Let M be a smooth, simply connected n-manifold with $n \geq 5$. If two automorphisms of M are pseudoisotopic (see Definition 1.22), then they are isotopic.*

Proof Suppose that $\phi, \phi' \in \mathrm{Diff}(M)$ are pseudoisotopic. Then $\psi := \phi' \circ \phi^{-1}$ is pseudoisotopic to Id_M. Let $\Psi\colon I \times M \to I \times M$ be the corresponding pseudoisotopy; that is, $\Psi(0,x) = x$ and $\Psi(1,x) = \psi(x)$ for $x \in M$.

Let $p\colon I \times M \to I$ be the projection, and consider the function

$$f := p \circ \Psi^{-1} \in C^\infty(I \times M).$$

Then f and p are Morse functions on $I \times M$ without critical points. Let $\{f_t : t \in I\}$ be a generic one-parameter family of smooth functions on $I \times M$ such that $f_0 = f$ and $f_1 = p$. Using Cerf theory, we can remove all the critical points of f_t for every $t \in I$, without changing f_0 and f_1.

Fix a Riemannian metric on $I \times M$. For $t \in I$, we define $\Psi_t \in \mathrm{Diff}(I \times M)$ by sending $(s,x) \in I \times M$ to the point obtained by flowing from $(0,x)$ along $\mathrm{grad}(f_t)$ for time $s \cdot m(t,x)$, where $m(t,x)$ is the time required to flow from $(0,x)$ to $\{1\} \times M$ along $\mathrm{grad}(f_t)$. Finally, let $\psi_t(x) := \Psi_t(1,x)$ for $x \in M$. Then $\psi_0 = \psi$ and $\psi_1 = \mathrm{Id}_M$, showing that ψ and Id_M are isotopic. This implies that ϕ and ϕ' are isotopic, as claimed. □

Remark 2.72 In fact, Cerf's argument also implies that any pseudoisotopy can be deformed to an isotopy.

Recall that Θ_n is the group of homotopy n-spheres that we defined in Section 2.8.1.

Corollary 2.73 *For $n \geq 6$, the following hold:*

(i) $\pi_0(\mathit{Diff}(D^n)) = 0$;
(ii) $\pi_0(\mathit{Diff}(S^{n-1})) \cong \Theta_n$;
(iii) $\pi_1(\mathit{Diff}(S^{n-1}))$ *is an extension of* Θ_{n+1}.

Proof To prove (i), let $\Phi \in \text{Diff}(D^n)$. Then we can isotope Φ such that it fixes the origin by composing with an isotopy of D^n taking $\Phi(0)$ to 0; see Corollary 1.38. Near 0, we can further isotope Φ such that $(d\Phi)_0 = \text{Id}_{T_0 D^n}$, and then such that $\Phi|_{D_0^n} = \text{Id}_{D_0^n}$ on a smaller disk $D_0^n \subset \text{Int}(D^n)$ about 0. Then $\Phi|_{\overline{D^n \setminus D_0^n}}$ is a pseudoisotopy of S^{n-1}, which we can isotope to the identity by Remark 2.72.

Parts (ii) and (iii) follow from the following exact sequence:

$$\pi_1(\text{Diff}(S^{n-1})) \to \pi_0(\text{Diff}(S^n)) \to \pi_0(\text{Diff}(D^n)) \to$$
$$\pi_0(\text{Diff}(S^{n-1})) \to \Theta_n \to 0.$$

Here, the first map assigns to a loop $\gamma \colon I \to \text{Diff}(S^{n-1})$ based at $\text{Id}_{S^{n-1}}$ the map obtained from

$$L \colon I \times S^{n-1} \to I \times S^{n-1}$$

defined as $L(t, \cdot) = \gamma(t)$ by capping off $\{0,1\} \times S^{n-1}$ with $\{0,1\} \times D^n$, and extending L identically. The second map is obtained by fixing an automorphism of S^n on the lower hemisphere using an isotopy as in the proof of (i), and restricting it to the upper hemisphere D^n of S^n. The third map is restriction to the equator $\partial D^n = S^{n-1}$. The last map assigns to an automorphism ϕ of S^{n-1} the homotopy sphere obtained by gluing two copies of D^n using the clutching function f; see Remark 1.56. If f extends to D^n, then the result is diffeomorphic to S^n. □

Cerf theory also plays a key role in the proof of Kirby's calculus that is one of the cornerstones of four-manifold topology; see Section 6.1.

We now list Cerf's moves. As earlier, $\{f_t : t \in I\}$ is a generic one-parameter family of smooth functions on an n-manifold M.

(i) **Independent Trajectories Principle**: Given two parts of a graphic, suppose that there are no gradient trajectories between the critical points in one part to those in the other part. Then we can deform one part relative to the other as in Figure 2.2.

(ii) **Uniqueness of birth**: If two consecutive critical points cancel and are immediately reborn, then we can change the graphic as in Figure 2.3, assuming that the level set where we add the pair of critical points is connected.

(iii) **Uniqueness of death**: Suppose that we have a cancelling pair of consecutive critical points c_1 and c_2 of f_t such that $f_t(c_1) < f_t(c_2)$, and they appear at $t - \varepsilon_1$ and disappear at $t + \varepsilon_2$. If $n \geq 6$ and the level set immediately

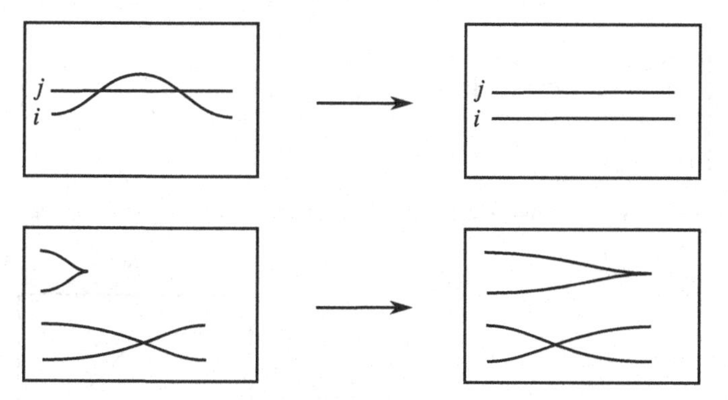

Figure 2.2 **Independent Trajectories Principle**: There are no gradient trajectories connecting the two parts involved; for example, $i < j$.

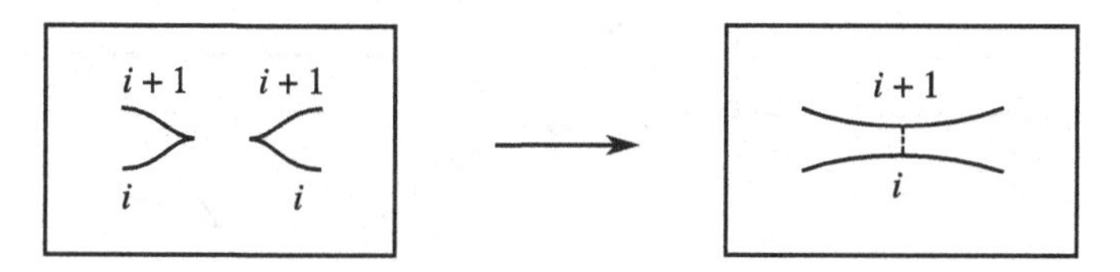

Figure 2.3 **Uniqueness of birth**: On the right, the critical points cancel.

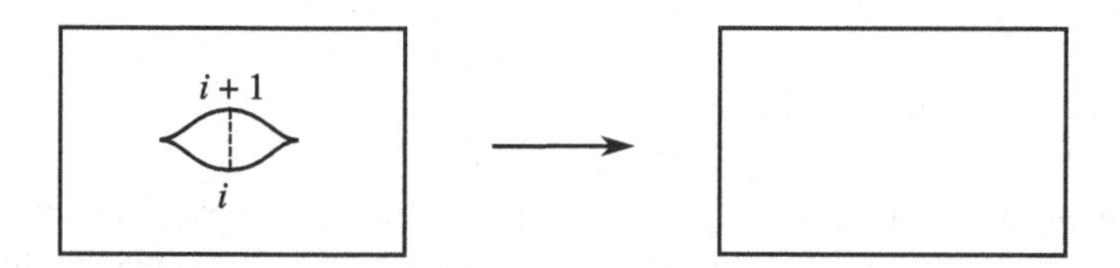

Figure 2.4 **Uniqueness of death**: On the left, the critical points cancel.

above c_1 (or, equivalently, immediately below c_2) is simply connected, then we can remove c_1 and c_2 from the graphic; see Figure 2.4.

(iv) **Triangle move**: Suppose that f_t has three consecutive critical points c_1, c_2, and c_3 with indices i_1, i_2, and i_3, respectively, such that $f(c_1) < f(c_2) < f(c_3)$, and at least one of the following conditions is satisfied:
(a) $i_1 + i_3 \leq n - 1$,
(b) $\min(i_1, i_3) \leq i_2 - 1$,
(c) $i_1 = i_2 = i_3 \leq n - 2$.
If the graphic near f_t looks as on the left of Figure 2.5, then we can deform the family to the graphic on the right. We can go from right to left if at least one of the following conditions is satisfied:
(a) $i_1 + i_3 \geq n + 1$,

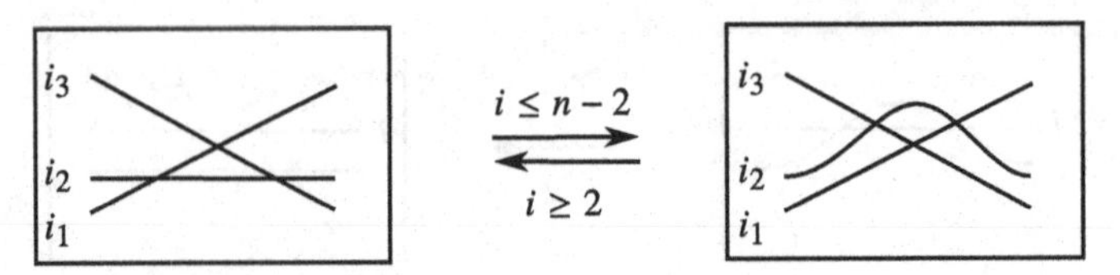

Figure 2.5 **Triangle move**: Here, we assume that $i_1 = i_2 = i_3$, which we denote by i.

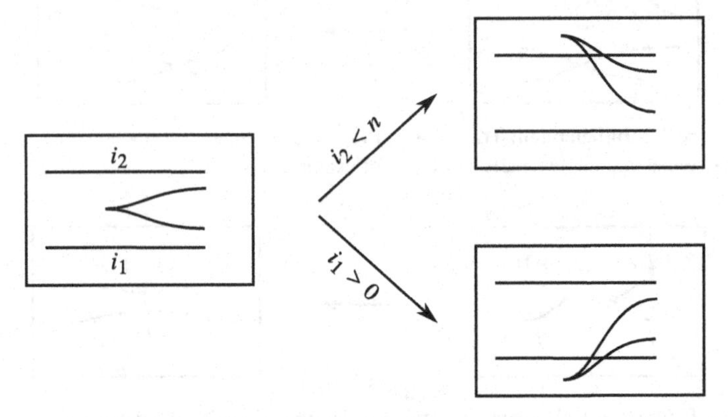

Figure 2.6 **Beak move**: We can perform the bottom move when $i_1 > 0$, and the top move when $i_2 < n$.

(b) $\min(i_1, i_3) \geq i_2 + 1$,
(c) $i_1 = i_2 = i_3 \geq 2$.

(v) **Beak move**: Suppose we have a portion of the Cerf graphic as on the left of Figure 2.6. If $i_1 > 0$, then we can deform the family to the one in the lower right. If $i_2 < n$, then we can transform it to the one in the upper right.

(vi) **Dovetail move**: Suppose we have the graphic in the top left of Figure 2.7. Let M_- be the level set immediately below c_1, and M_+ the level set immediately above c_1. Suppose $n \geq 6$, and either $0 \leq i \leq n-4$ and $\pi_1(M_-) = 0$, or $i = n-3$ and $\pi_1(M_+) = 0$. Then we can deform the family to the graphic in the top right. We obtain the move in the bottom row of the figure by replacing f with $-f$. When $\operatorname{ind}(c_3) = 0$ and we are in the top row, or $\operatorname{ind}(c_3) = n$ and we are in the bottom row, we do not need any additional assumptions.

The preceding moves are all reversible, except we need the additional restrictions mentioned for the Triangle move. Furthermore, we do not need any dimension or simple connectivity assumptions for the reverse of Uniqueness of death.

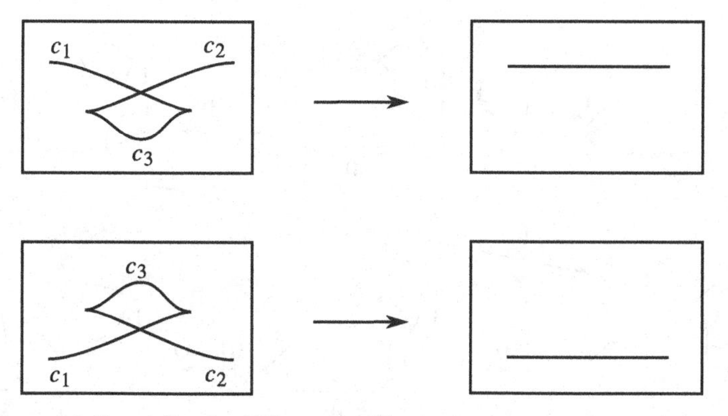

Figure 2.7 **Dovetail move**: This works without extra assumptions if $\operatorname{ind}(c_3) = 0$ in the top row, and when $\operatorname{ind}(c_3) = n$ in the bottom row.

The reason behind the term 'Uniqueness of birth' is the following: The one-parameter families $\{f_{(1-t)/2} : t \in I\}$ and $\{f_{(1+t)/2} : t \in I\}$ are both births starting from $f_{1/2}$. Any two families involving a single birth are equivalent, so their endpoints f_0 and f_1 are also equivalent Morse functions. Hence, f_0 and f_1 can be connected by a family of Morse functions with no crossings between critical values. Uniqueness of death has an analogous interpretation.

Definition 2.74 We say that a Morse function f is *ordered* if $\operatorname{ind}(p) < \operatorname{ind}(q)$ for $p, q \in \operatorname{Crit}(f)$ implies that $f(p) < f(q)$.

Lemma 2.75 *Any two ordered Morse functions f_0 and f_1 can be connected by a generic one-parameter family $\{f_t : t \in I\}$ of smooth functions such that the Cerf graphic has finitely many births, possibly followed by crossings among critical points of the same index, and finally finitely many deaths. Furthermore, whenever f_t is Morse, it is ordered.*

Proof This can be achieved using the Independent Trajectories Principle and the Beak move. □

We can further strengthen Lemma 2.75 as follows:

Lemma 2.76 *Let f_0 and f_1 be ordered Morse functions on the n-manifold M that both have zero or one index zero (or n) critical points. Then they can be connected by a generic one-parameter family $\{f_t : t \in I\}$ of smooth functions such that the number of index zero (or n) critical points is constant in t.*

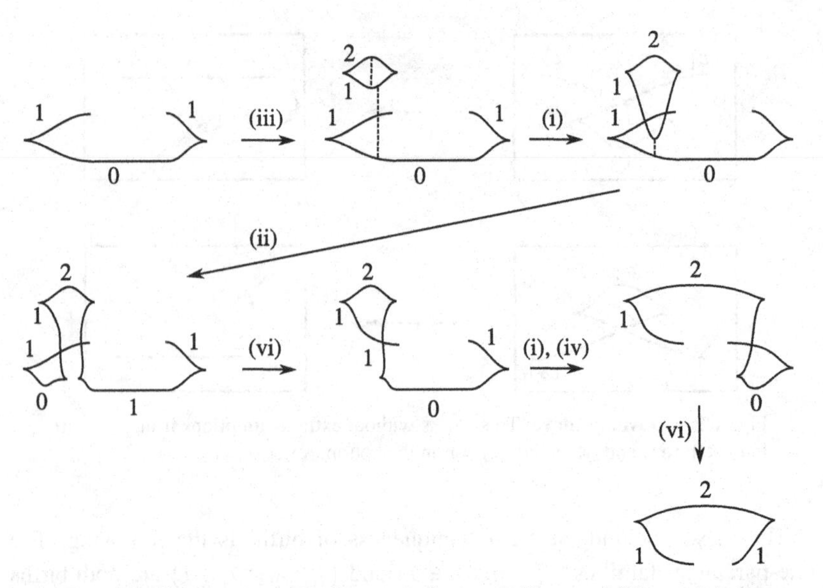

Figure 2.8 The sequence of Cerf moves in the proof of Lemma 2.76 to trade index zero critical points born with index one and two critical points. The horizontal arrows are labelled by the number of the corresponding Cerf move. There might be other index one critical points that we do not show.

When $n > 2$, and f_0 and f_1 both have zero or one index zero and n critical points, then we can assume the number of index zero and n critical points are both constant.

Proof The main idea is that we can trade all the index zero critical points born with index one and two critical points. (Compare this with how we traded index one critical points with index three critical points in the proof of the h-cobordism theorem.) More specifically, consider the first index zero critical point born, and call it p_t. This exists until it gets cancelled with an index one critical point s_t. Using the reverse of Uniqueness of death, we add a cancelling pair of index one and two critical points q_t and r_t such that the index one critical point q_t also cancels with p_t. For a schematic picture, see Figure 2.8. We then push down q_t using the Independent Trajectories Principle, the Beak move, until it lies just above p_t. We then cancel p_t and q_t using Uniqueness of birth, and remove the resulting dovetail involving p_t. We push the remaining second branch of the Cerf graphic containing q_t using the Independent Trajectories Principle, the Beak move, and the Triangle move to the end, and remove the resulting dovetail including q_t and p_t. This way, we have completely eliminated p_t and only introduced index one and two critical points.

The claim about index n critical points is analogous. □

3
Three-Manifolds

Every topological three-manifold admits a unique smooth structure up to diffeomorphism by the work of Moise [120], hence the categories of topological and smooth manifolds are equivalent in dimension three.

The classification of closed, orientable three-manifolds is governed by the geometric structures that they admit, which was conjectured by Thurston [170] and proven by Perelman. For a complete proof of Thurston's geometrisation conjecture, see, for example, Morgan and Tian [122]. In particular, this work settled the century-old Poincaré conjecture.

To better understand Thurston's geometrisation programme, we first look at constant Gaussian curvature metrics on closed, orientable surfaces: S^2 admits a metric of constant curvature one, T^2 a flat metric, and every surface of genus greater than one admits a hyperbolic metric, which has curvature -1. Recall that, by the Gauss–Bonnet theorem, the integral of the Gaussian curvature of a surface S is $2\pi\chi(S)$. So, a constant curvature κ metric on S^2 has $\kappa > 0$, on T^2 it is flat, and on every other closed, orientable surface it has $\kappa < 0$.

Dimension three is more complicated but still closely tied to Riemannian geometry. A three-manifold is prime if every separating embedded sphere bounds a ball. Every closed, orientable three-manifold can be uniquely written as a connected sum of prime three-manifolds, up to permuting the components and adding three-spheres. Prime pieces can further be decomposed along tori such that each piece is atoroidal (i.e., contains no incompressible torus) or Seifert fibred (i.e., admits a circle fibration that has some singular fibres of a prescribed form). By the JSJ decomposition theorem, such a torus decomposition exists and is essentially unique if the number of tori is minimal.

While not every closed, orientable three-manifold admits a nice geometry, they can always be cut along spheres and tori, as in the previous paragraph, such that each piece does admit one of eight model geometries. In addition to Euclidean, spherical, and hyperbolic, the model geometries consist of $S^2 \times \mathbb{R}$,

$H^2 \times \mathbb{R}$, the universal cover of $\mathrm{SL}(2,\mathbb{R})$, and Nil and Sol; see Scott [160]. Except for the hyperbolic and Sol pieces, all other geometric pieces are Seifert fibred. The proof is not topological but uses the Ricci flow and techniques from PDEs and geometric analysis. The basic idea is that one chooses an arbitrary Riemannian metric on the three-manifold, which then evolves in a way that evens out the Ricci curvature, like a heat flow. This flow develops singularities in finite time, at which point a surgery can be carried out that corresponds to cutting along a sphere or a torus. The flow can be extended further after the surgery. In the end, one obtains pieces where the metric limits to one of the eight model geometries.

As Sol and Seifert fibred three-manifolds are easier to study, the most interesting class is that of hyperbolic three-manifolds. By Mostow–Prasad rigidity [123][146], the hyperbolic structure is unique, so geometric invariants, such as the volume, are also topological invariants.

In this book, we only discuss classical results from three-manifold topology, preceding Thurston's revolutionary work. Most of the proofs use cut-and-paste techniques. Some of these results will then be applied in the following section on knots and links.

After proving the Schönflies theorem in dimension three, we introduce incompressible surfaces and prove the celebrated loop theorem. We then discuss Haken manifolds, which form a class of three-manifolds for which one can perform inductive proofs by cutting them into simpler pieces along incompressible surfaces, and that are determined by their fundamental groups. We develop normal surface theory in order to show that every three-manifold has an essentially unique prime decomposition with respect to connected sum. Closed three-manifolds can be described by gluing together two handlebodies, which is called a Heegaard decomposition. If the handlebodies are solid tori and the manifold is not S^3 or $S^1 \times S^2$, then we call the resulting manifold a lens space. We conclude this chapter by classifying lens spaces using Reidemeister torsion.

3.1 The Schönflies Theorem

The Jordan–Schönflies theorem states that every Jordan curve in $\mathbb{R}^2$ bounds a disk. One might wonder whether this generalises to higher dimensions: does an embedding of an n-sphere into $\mathbb{R}^{n+1}$ bound a disk? The answer depends on the category we are working in. The Alexander horned sphere is an example of a continuous embedding of D^3 into $\mathbb{R}^3$ whose complement has a non-finitely generated fundamental group; see Rolfsen [155] for the construction. Its boundary is hence a continuously embedded S^2 in S^3 that does

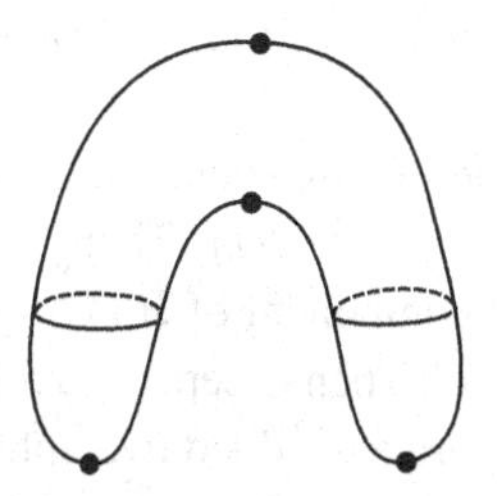
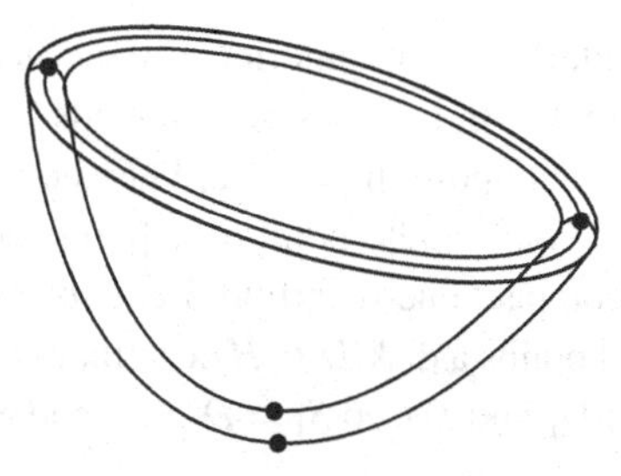

Figure 3.1 Two embeddings of S^2 in $\mathbb{R}^3$ with one saddle point and two local minima.

not bound a disk on one side. If we remove a point from S^3 that lies in the D^3 component, we obtain a continuously embedded S^2 in $\mathbb{R}^3$ that does not bound a disk.

However, if we restrict to the smooth or PL category, then the generalised Schönflies theorem is known to hold for $n \neq 3$: a smoothly or PL embedded S^n bounds a D^{n+1} in $\mathbb{R}^{n+1}$. In the smooth category, for $n > 3$ this follows from the h-cobordism theorem; see Proposition 2.8. The case $n = 3$ is still open. The case $n = 2$ is due to Alexander, which we prove in the smooth category:

Theorem 3.1 (Schönflies theorem) *Any smoothly embedded two-sphere S in $\mathbb{R}^3$ bounds a three-disk.*

Proof We follow the exposition of Casson [16]. We use the following observation: if $M = M_1 \cup M_2$, where M_1 is a three-manifold, $M_2 \approx D^3$, and

$$M_1 \cap M_2 = \partial M_1 \cap \partial M_2 \approx D^2,$$

then $M \approx M_1$.

We can perturb S such that the height function $x_3|_S$ is Morse for $x_3 \colon \mathbb{R}^3 \to \mathbb{R}$. Let M be the closure of the bounded component of $\mathbb{R}^3 \setminus S$. The proof proceeds by induction on the number n of index 1 critical points of $x_3|_S$.

If $n = 0$, then S has one local minimum and one local maximum. Let $t_{\min} := \min(x_3|_S)$ and $t_{\max} := \max(x_3|_S)$. Then, for every $t \in (t_{\min}, t_{\max})$, the level set $(x_3|_S)^{-1}(t)$ is a circle in $\mathbb{R}^2 \times \{t\}$ that bounds a two-disk in $\mathbb{R}^2 \times \{t\}$ by the smooth Jordan–Schönflies theorem, and hence $M \approx D^3$.

If $n = 1$, then there are either two local minima and one local maximum, or one local minimum and two local maxima. Without loss of generality, assume it is the former. Consider the level sets of $x_3|_S$. As we pass the two local minima, two circles appear. If these have disjoint interiors, then M is a regular neighbourhood of a $\cap$-shaped curve; see the left of Figure 3.1. If one circle is contained in the interior of the other, M is a 'tilted bowl'; see the right of Figure 3.1. In both cases, M bounds a disk.

Now suppose that $n \geq 2$. Let t be a regular value of $x_3|_S$ such that there is at least one saddle point of $x_3|_S$ on each side of $H := \mathbb{R}^2 \times \{t\}$. Then $S \cap H$ is a compact one-manifold. Let C be an innermost component of $H \cap S$ in H. This bounds a disk $D \subset H$ such that $D \cap S = \partial D = C$. Then C separates S into disks D_1 and D_2, so $S_1 = D \cup D_1$ and $S_2 = D \cup D_2$ are embedded two-spheres in $\mathbb{R}^3$.

If both $x_3|_{S_1}$ and $x_3|_{S_2}$ have saddles, then they bound disks M_1 and M_2, respectively, by the inductive hypothesis. There are three cases: If $M_1 \cap M_2 = D$, then $M = M_1 \cup M_2 \approx D^3$ by the observation. If $M_1 \subset M_2$, then we can apply the observation to $M = M_1 \cup (\overline{M_2 \setminus M_1})$ as $M_1 \cap (\overline{M_2 \setminus M_1}) = D_1$ is a disk. The case $M_2 \subset M_1$ is analogous.

If S_1 has no saddle, it bounds a disk M_1 by the $n = 0$ case. As before, using the observation, S bounds a disk if and only if S_2 bounds a disk. We push S_2 a bit to eliminate C from $S_2 \cap H$, reducing the number of its components. We repeat the preceding procedure with a new innermost circle, which has to terminate since there is a saddle on each side of H. It follows that S also bounds a disk. □

Exercise 3.2 Using the same method as in the proof of the Schönflies theorem (and assuming the Schönflies theorem), show that if T is a smooth torus in S^3, then one of the components of $S^3 \setminus T$ has closure $S^1 \times D^2$. Give an example of a torus T in $\mathbb{R}^3$ such that the closure of the bounded component of $\mathbb{R}^3 \setminus T$ is not a solid torus.

3.2 Incompressible Surfaces and the Loop Theorem

Definition 3.3 We say that the surface S in the three-manifold M with boundary is *properly embedded* if $S \cap \partial M = \partial S$, and S is transverse to ∂M.

Definition 3.4 Let S be an embedded surface in a three-manifold M such that $\partial S \subset \partial M$. Then we say that S is *compressible* if either there is a disk $D \subset M$ such that $D \cap S = \partial D$ and ∂D does not bound a disk in S, or if S has an S^2 component that is the boundary of a three-ball in M, or if S has a boundary-parallel D^2 component. We say that S is *incompressible* otherwise. We call such a D a *compressing disk.*

Given a compressing disk D for a surface S in the three-manifold M, we can compress it along D as follows: Let $N(D) \approx D \times [-1, 1]$ be a regular neighbourhood of D such that $N(D) \cap S \approx \partial D \times [-1, 1]$. Then the compressed surface is

$$S(D) := S \setminus (\partial D \times (-1, 1)) \cup (D \times \{-1, 1\}).$$

One can think of this as a surgery operation for embedded surfaces.

Definition 3.5 A surface S in a three-manifold M is called *two-sided* if its normal bundle is trivial. We say that S is *π_1-injective* if the map $\pi_1(S) \to \pi_1(M)$ induced by the inclusion is injective.

For example, if both M and S are orientable, then S is two-sided. The following fundamental result was first stated by Dehn in 1910, but gaps in his proof were found by Kneser. The first complete proof was given by Papakyriakopoulos [144] in 1957.

Theorem 3.6 (Dehn's lemma) *Let M be a three-manifold with boundary, and $f\colon D^2 \to M$ a continuous map that is an embedding near ∂D^2 and $f(\partial D^2) \subset \partial M$. Then there exists an embedding $g\colon D^2 \to M$ such that f and g agree in a neighbourhood of ∂D^2.*

The following generalisation of Dehn's lemma is known as the loop theorem, also due to Papakyriakopoulos:

Theorem 3.7 (loop theorem) *Let M be a three-manifold with boundary (not necessarily compact). If ∂M is not π_1-injective, then it is compressible.*

Proof We follow the exposition of Hatcher [58]. Let

$$f\colon (D^2, \partial D^2) \to (M, \partial M)$$

be a continuous map such that $[f|_{\partial D^2}] \neq 1 \in \pi_1(\partial M)$. Choose a triangulation of M. By the relative version of the simplicial approximation theorem, there is a triangulation of D^2 such that f is homotopic to a simplicial map $f_0\colon (D^2, \partial D^2) \to (M, \partial M)$. Let $D_0 := f_0(D^2)$, and V_0 a regular neighbourhood of D_0 obtained by taking the union of the simplices of the second barycentric subdivision of M incident to D_0.

Let $p_1\colon M_1 \to V_0$ be a connected double cover of V_0, if one exists. See Figure 3.2. As $\pi_1(D^2) = 1$, we can lift f_0 to a map $f_1\colon D^2 \to M_1$. We write $D_1 := f_1(D^2)$ and V_1 for a regular neighbourhood of D_1, as earlier. We repeat this procedure to obtain a sequence of maps $f_i\colon D^2 \to M_i$ whose image is D_i with regular neighbourhood V_i, where $p_i\colon M_i \to V_{i-1}$ is a connected double

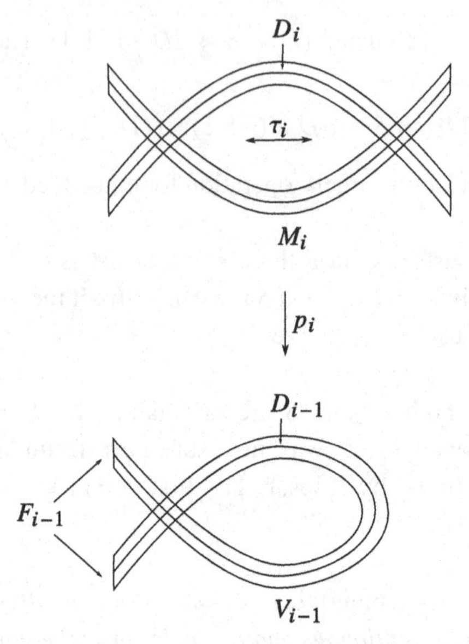

Figure 3.2 The double cover $p_i : M_i \to V_{i-1}$ in the proof of the loop theorem. Here, V_{i-1} is a regular neighbourhood of the disk D_{i-1} and τ_i is the covering involution.

cover. We can triangulate M_i by lifting the triangulation of V_{i-1}, and D_i and V_i are subcomplexes.

We claim that this process terminates in finitely many steps, and write n for the largest index for which V_n admits no connected double cover. Indeed, let

$$E_i := p_i^{-1}(D_{i-1}),$$

and note that the double cover $E_i \to D_{i-1}$ is connected as M_i is connected and deformation retracts onto E_i by lifting the retraction of V_{i-1} onto D_{i-1}. If τ_i is the covering automorphism of $E_i \to D_{i-1}$, then $E_i = D_i \cup \tau_i(D_i)$. As E_i is connected, $D_i \cap \tau_i(D_i) \neq \emptyset$. So, there is a simplex σ of D_i such that $\tau_i(\sigma) \subset D_i$. As τ_i has no fixed points, $\sigma \neq \tau_i(\sigma)$. This means that D_{i-1}, which is a quotient of D_i, has fewer simplices than D_i. The number of simplices in each D_i is bounded from above by the number of simplices in D^2, which hence gives an upper bound on the number of steps n.

We now show that each component of ∂V_n is a two-sphere. It suffices to prove that

$$H_1(\partial V_n; \mathbb{Z}_2) = 0.$$

As V_n has no connected double cover, $\pi_1(V_n)$ has no index two subgroup, so

$$H^1(V_n; \mathbb{Z}_2) \cong \mathrm{Hom}(\pi_1(V_n), \mathbb{Z}_2) = 0.$$

So, by the universal coefficient theorem, we have $H_1(V_n;\mathbb{Z}_2) = 0$, and, by Poincaré duality, $H_2(V_n,\partial V_n;\mathbb{Z}_2) = 0$. When these are combined with the exact sequence

$$H_2(V_n,\partial V_n;\mathbb{Z}_2) \to H_1(\partial V_n;\mathbb{Z}_2) \to H_1(V_n;\mathbb{Z}_2)$$

of the pair $(V_n,\partial V_n)$, we see that $H_1(\partial V_n;\mathbb{Z}_2) = 0$, as claimed.

Let $\partial_0 V_i$ be the component of ∂V_i containing $f_i(\partial D^2)$, and let

$$F_i := (p_1 \circ \cdots \circ p_i)^{-1}(\partial M) \cap \partial_0 V_i.$$

We denote the kernel of the homomorphism

$$(p_1 \circ \cdots \circ p_i)_* \colon \pi_1(F_i) \to \pi_1(\partial M)$$

by N_i. As $[f|_{\partial D^2}] \neq 0 \in \pi_1(\partial M)$, we have $[f_i|_{\partial D^2}] \notin N_i$.

Since each component of ∂V_n is a sphere, the surface F_n is planar. Hence $\pi_1(F_n)$ is normally generated by the components of ∂F_n. As $N_n \neq \pi_1(F_n)$, there is a component C of ∂F_n that represents an element of $\pi_1(F_n) \setminus N_n$. Each component of $\partial_0 V_n \setminus F_n$ is a disk. We can push the interior of the disk bounding C into V_n to obtain a smooth embedding $g_n \colon D^2 \hookrightarrow V_n$ such that $[g_n|_{\partial D^2}] \notin N_n$.

Starting from g_n, we recursively construct a sequence of embeddings

$$g_i \colon D^2 \hookrightarrow V_i$$

with $[g_i|_{\partial D^2}] \notin N_i$. Then g_0 is the desired embedding of D^2 into M whose boundary is homotopically non-trivial.

Suppose we have already constructed g_i. To obtain g_{i-1}, we consider the immersion $p_i \circ g_i$. After a small perturbation of g_i, this is an immersion with transverse double point curves and arcs, since p_i is a two-fold cover. We use cut-and-paste techniques to get rid of these, resulting in the embedding g_{i-1}.

Let C be a double point curve of $p_i \circ g_i(D^2)$. Let $N(C)$ be a regular neighbourhood of C in $p_i \circ g_i(D^2)$. Then $N(C)$ is an X-bundle over S^1. This can be obtained from $X \times [0,1]$ by identifying $X \times \{0\}$ and $X \times \{1\}$ via an automorphism φ of the figure X. This automorphism cannot be a 90° rotation or a reflection in one of the two lines in X as otherwise D^2 would contain a Möbius band. So, φ is either the identity of the figure X or a reflection in a vertical or horizontal line.

First, suppose that φ is the identity. Then the pre-image of C in D^2 consists of two circles that bound disks D and D'. If these disks are nested, say $D \subset D'$, then we replace $p_i \circ g_i|_{D'}$ with $p_i \circ g_i|_D$ and smooth the corners; see the top left of Figure 3.3. If $D \cap D' = \emptyset$, then we swap $p_i \circ g_i|_D$ and $p_i \circ g_i|_{D'}$ and

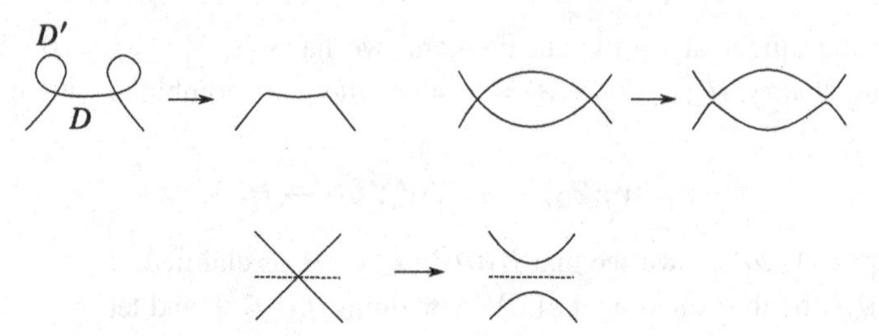

Figure 3.3 Removing double point curves of an immersed disk. Top left: the pre-images of the double point curve are nested circles. Top right: the pre-images of the double point curve are not nested circles. Bottom: the pre-images of the double point curve is connected.

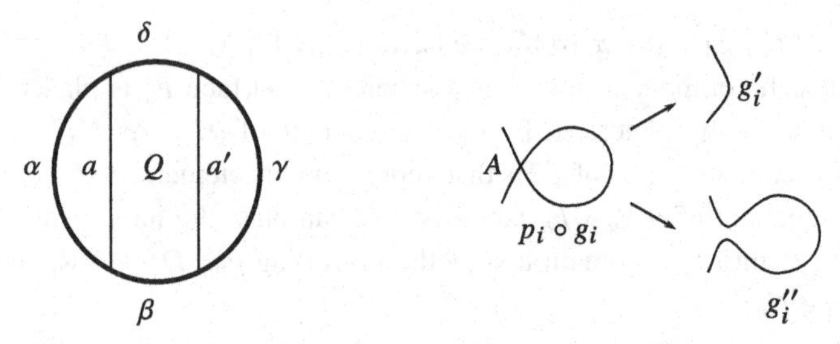

Figure 3.4 Left: The pre-images of a double point arc A. Right: The two resolutions g_i' and g_i''.

again smooth the corners; see the top right of Figure 3.3. If φ is a reflection in the horizontal axis, we smooth each X-fibre as on the bottom of Figure 3.3, which replaces the annulus mapping to $N(C)$ with another annulus. The case of reflection in a vertical axis is analogous. In each case, we remove C from the double point set, without introducing any new double point curves or changing $p_i \circ g_i|_{\partial D^2}$.

Now suppose that A is a double point arc of $p_i \circ g_i(D^2)$. Its pre-image in D^2 consists of a pair of arcs a, $a' \subset D^2$ with boundary on ∂D^2; see the left of Figure 3.4. We label the components of $\partial D^2 \setminus (a \cup a')$ by α, β, γ, and δ counterclockwise such that a and α and a' and γ form bigons B and B' (and hence a, β, a', δ are sides of a quadrilateral Q).

There are two ways of smoothing the double point arc A; see the right of Figure 3.4. In the first case, we glue together the maps $p_i \circ g_i|_B$ and $p_i \circ g_i|_{B'}$, smooth the corners, and remove the quadrilateral Q to obtain the map g_i'. The other smoothing of A gives rise to a map we denote g_i''. This is obtained by removing $p_i \circ g_i|_Q$ and vertically reversing it, gluing the side a of B to the side a' of Q, and the side a' of B' to the side a of Q.

We claim that either $[g_i'|_{\partial D^2}] \notin N_{i-1}$ or $[g_i''|_{\partial D^2}] \notin N_{i-1}$. Choose an orientation of A; this induces orientations on a and a'. If exactly one of them is oriented coherently with ∂Q, then

$$\begin{aligned}\alpha\beta\gamma\delta &= (\alpha\gamma)\delta^{-1}(\alpha\beta^{-1}\gamma\delta^{-1})^{-1}(\alpha\gamma)\delta \\ &= (g_i'|_{\partial D^2})\delta^{-1}(g_i''|_{\partial D^2})^{-1}(g_i'|_{\partial D^2})\delta.\end{aligned}$$

Otherwise, we have

$$\begin{aligned}\alpha\beta\gamma\delta &= (\alpha\gamma^{-1})(\gamma\delta)^{-1}(\alpha\gamma^{-1})^{-1}(\alpha\delta\gamma\beta)(\gamma\delta) \\ &= (g_i'|_{\partial D^2})(\gamma\delta)^{-1}(g_i'|_{\partial D^2})^{-1}(g_i''|_{\partial D^2})(\gamma\delta).\end{aligned}$$

As $\alpha\beta\gamma\delta \notin N_{i-1}$, at least one of $[g_i'|_{\partial D^2}]$ and $[g_i''|_{\partial D^2}]$ is not in N_{i-1}, as claimed, and we continue with this map until $p_i \circ g_i$ becomes an embedding. □

It is clear that every π_1-injective surface without S^2 components that bound three-balls and without boundary-parallel D^2 components is incompressible. The converse also holds for two-sided surfaces:

Theorem 3.8 *If S is a two-sided, incompressible surface in a three-manifold M, then S is π_1-injective.*

Proof By contradiction, suppose that there is a homotopically non-trivial curve $\gamma\colon S^1 \to S$ that is null-homotopic in M. Then there is a smooth map $u\colon D^2 \to M$ transverse to S such that $u|_{S^1} = \gamma$. The pre-image $u^{-1}(S)$ is a smooth one-manifold C in D^2. Let C_0 be an innermost component of C. Let $D_0 \subset D^2$ be the disk bounded by C_0. If $u|_{C_0}$ is null-homotopic in S, then we can replace u on D_0 with the null-homotopy and push it off S such that we obtain a map v with $v^{-1}(S) = C \setminus C_0$. Repeating this procedure, we can assume that $u|_{C_0}$ is not null-homotopic. Note that $u(D_0) \cap S = u(C_0)$. Hence, we can assume the curve γ on S is null-homotopic in the complement of S.

As S is two-sided, $N(S) \approx S \times [-1, 1]$. The curve $\gamma \times \{1\}$ is null-homotopic in the complement

$$M' := M \setminus (S \times (-1, 1)).$$

Hence, by Theorem 3.7 applied to M', the surface $S \times \{1\}$ is compressible in M', so S is compressible in M, which is a contradiction. □

Exercise 3.9 Let K denote the Klein bottle, and let $K\widetilde{\times}I$ be an orientable I-bundle over K. Then $\partial(K\widetilde{\times}I)$ is a two-torus T. Show that we can attach $S^1 \times D^2$ to $K\widetilde{\times}I$ along T such that in the resulting three-manifold M, the Klein bottle K is incompressible but not π_1-injective.

Papakyriakopoulos also proved the sphere theorem, whose proof can be found, for example, in Hatcher [58]:

Theorem 3.10 (sphere theorem) *Let M be a connected three-manifold with $\pi_2(M) \neq 0$. Then either there is an embedded two-sphere in M representing a non-trivial element of $\pi_2(M)$ or there is an embedded two-sided $\mathbb{RP}^2$ in M such that the composition of the cover $S^2 \to \mathbb{RP}^2$ with the inclusion $\mathbb{RP}^2 \hookrightarrow M$ represents a non-trivial element of $\pi_2(M)$.*

3.3 Haken Manifolds

In this section, we review some results on a class of particularly nice three-manifolds called Haken manifolds that includes knot exteriors. They were used by Haken in his unknot detection algorithm.

Definition 3.11 A three-manifold M is *irreducible* if every embedded two-sphere in M bounds a three-ball. We say that M is *boundary-irreducible* if each component of ∂M is incompressible in M.

An oriented three-manifold M is *Haken* or *sufficiently large* if it is irreducible, and contains a properly embedded, orientable, incompressible surface.

We will need a result that is known as the 'half lives, half dies lemma'.

Lemma 3.12 *Let M be a compact, oriented three-manifold. Then*

$$rk\,(\ker(H_1(\partial M) \to H_1(M))) = \frac{rk\,(H_1(\partial M))}{2}.$$

Proof We follow Hatcher [58, lemma 3.5]. From the homological and cohomological long exact sequences of the pair $(M, \partial M)$ with real coefficients, we obtain the commutative diagram

$$\begin{array}{ccccc} H_2(M,\partial M;\mathbb{R}) & \xrightarrow{\partial_*} & H_1(\partial M;\mathbb{R}) & \xrightarrow{i_*} & H_1(M;\mathbb{R}) \\ \downarrow \approx & & \downarrow \approx & & \downarrow \approx \\ H^1(M;\mathbb{R}) & \xrightarrow{i^*} & H^1(\partial M;\mathbb{R}) & \xrightarrow{\delta} & H^2(M,\partial M;\mathbb{R}), \end{array}$$

where the vertical arrows are Poincaré duality isomorphisms. Hence

$$\begin{aligned} \dim \operatorname{coker}(i^*) &= \dim \operatorname{coker}(\partial_*) = \dim H_1(\partial M;\mathbb{R}) - \dim \operatorname{Im}(\partial_*) \\ &= \dim H_1(\partial M;\mathbb{R}) - \dim \ker(i_*). \end{aligned} \tag{3.1}$$

Since we are working over the field $\mathbb{R}$, we have $i^* = \mathrm{Hom}(i_*, \mathbb{R})$, and hence

$$\dim \ker(i_*) = \dim \operatorname{coker}(i^*).$$

Together with Equation (3.1), we obtain that

$$2 \dim \ker(i_*) = \dim H_1(\partial M; \mathbb{R}),$$

as claimed. □

Proposition 3.13 *Let M be a compact, oriented, irreducible three-manifold with boundary and $b_1(M) > 0$. Then M is Haken.*

Proof Since $b_1(M) > 0$, we have

$$H^1(M) \cong \mathrm{Hom}(H_1(M), \mathbb{Z}) \neq 0.$$

If $a \in H^1(M) \setminus \{0\}$, then there is a map $f\colon M \to S^1$ such that $f^*(1) = a$, where 1 is the generator of $H^1(S^1)$ dual to the fundamental class of S^1, using the identification between $H^1(M)$ and the set of homotopy classes $[M, S^1]$. Let x be a regular value of f. Then $S := f^{-1}(x)$ is Poincaré dual to a. It is orientable since we can pull back $\partial/\partial\theta \in T_x S^1$ to give a trivialisation of the normal bundle of S, and M is orientable. After compressing S, we obtain a properly embedded, orientable, incompressible surface S' homologous to S. Since M is irreducible, every S^2 component of S' bounds a ball. So we can remove these and all boundary-parallel D^2 components to obtain a surface S'' without changing the homology class. As $[S''] \neq 0$, we have $S'' \neq \emptyset$. So S'' is incompressible, and hence M is Haken. □

A Haken manifold M admits a *Haken hierarchy*, where we successively cut it along incompressible surfaces until we obtain three-balls. This makes them amenable to inductive proofs. Let S be a properly embedded, orientable, incompressible surface in M. Then we cut M along S; that is, consider $M_1 := M \setminus N(S)$. If a component M_1' of M_1 has a boundary component that is not a sphere, it has $b_1(M_1') > 0$ by the half lives, half dies lemma (Lemma 3.12), and is hence also Haken by Proposition 3.13. So we can cut M_1 along an incompressible surface. We can continue this procedure until all boundary components are spheres. As M was irreducible, each such sphere bounds a ball in M, and so all the resulting components are balls. The fact that this procedure terminates in finitely many steps relies on Theorem 3.23; see Hempel [62, theorem 13.3].

According to the following result of Waldhausen [175], Haken manifolds are determined by their fundamental groups:

Theorem 3.14 *Let M and N be compact, orientable three-manifolds that are irreducible and boundary-irreducible. Suppose M is Haken. Let $\psi\colon \pi_1(N) \to \pi_1(M)$ be an isomorphism that 'respects the peripheral structure'; that is, for each component F of ∂N there exists a component G of ∂M such that $\psi(i_*(\pi_1(F))) \subset A$ and A is conjugate in $\pi_1(M)$ to $i_*(\pi_1(G))$, where the i_* denote inclusion homomorphisms. Then there exists a homeomorphism $f\colon N \to M$ that induces ψ.*

The following deep result is due to Agol [4], building on the work of Kahn and Marković [75][76], which was known as the virtually Haken conjecture:

Theorem 3.15 *Every compact, orientable, irreducible three-manifold with an infinite fundamental group is virtually Haken; that is, it has a finite cover that is Haken.*

3.4 Normal Surfaces and Prime Decomposition

This section is based on unpublished notes of Casson [16].

Definition 3.16 We say that the closed three-manifold M is *prime* if, whenever $M = A \# B$, we have $A \approx M$ and $B \approx S^3$, or $A \approx S^3$ and $B \approx M$.

Every irreducible three-manifold is clearly prime. Conversely, we have the following:

Exercise 3.17 Show that if a closed three-manifold is prime, then it is either irreducible, or $S^1 \times S^2$, or the non-orientable S^2-bundle over S^1.

Normal surfaces were introduced by Kneser [88] in order to prove that every closed three-manifold can be written as a connected sum of prime three-manifolds, and further developed by Haken [54] into what is now known as normal surface theory. It is a fundamental tool for many three-manifold algorithms, including the following result due to Haken [55], Waldhausen [175], Jaco–Shalen [68], Johannson [70], and Hemion [61]:

Theorem 3.18 *There is an algorithm to decide whether two Haken three-manifolds are homeomorphic.*

While most of this book focuses on differential topological methods, normal surface theory is specific to the PL category of triangulated three-manifolds.

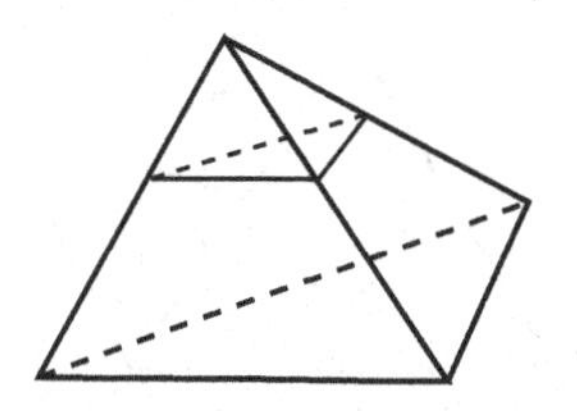
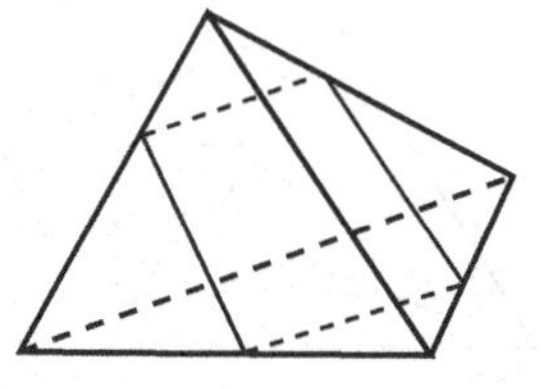

Figure 3.5 A normal surface intersects each tetrahedron in triangles (left) or quads (right).

In dimension three, every topological three-manifold has a unique smooth structure and a unique PL structure, so the two approaches are in some sense equivalent.

Let M be a three-manifold with triangulation T. We write T^i for the i-skeleton of T. Let F be a closed surface in M, not necessarily connected. We can make F transverse to T^i for every i.

Definition 3.19 We say that the surface F in the three-manifold M with triangulation T is *normal* if, for every three-simplex τ of T, each component of $F \cap \tau$ is either

(i) a *triangle* with vertices on three edges meeting at a vertex of τ, one vertex on each edge, as on the left of Figure 3.5, or
(ii) a quadrilateral (*quad*) with vertices on four edges that are adjacent to an edge of τ, one on each edge, as on the right of Figure 3.5.

There are four types of triangles, corresponding to the four vertices of τ, and three types of quads, corresponding to disjoint pairs of edges of τ. For each τ, there might be several components of $F \cap \tau$ of each triangle type, but at most one type of quad, as otherwise F would have double points.

Definition 3.20 Let M be a closed three-manifold and $S \subset M$ be a union of two-spheres. We say that these spheres are *independent* if no component of $M \setminus S$ is homeomorphic to $S^3 \setminus P$ for $P \subset S^3$ finite.

Lemma 3.21 *Let S be a surface embedded in the three-manifold M with triangulation T.*

(i) *If S is incompressible and M is irreducible, then S is isotopic to a normal surface.*
(ii) *If M is closed and S is a collection of k independent two-spheres, then M also contains an independent set of k two-spheres that is normal with respect to T.*

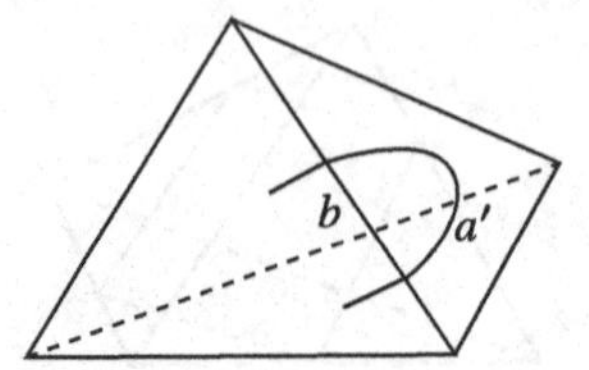

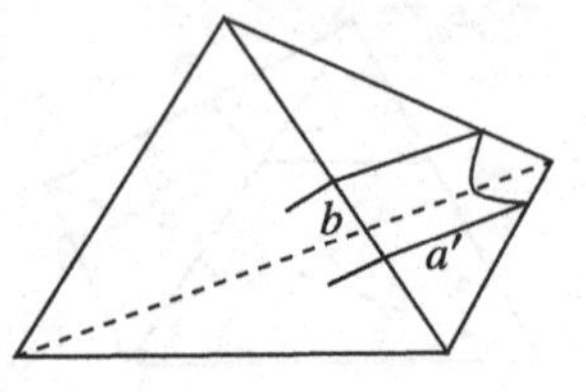

Figure 3.6 A component C of $S \cap \partial\Delta$ that is not of the required form admits one of these configurations.

Proof We first prove statement (i). We perturb S such that it becomes transverse to all simplices of T. We then define the *weight* of S to be $w(S) := |S \cap T^1|$. Suppose that S minimises $w(S)$ in its isotopy class. Then we claim that, for each three-simplex Δ of T, every component C of $S \cap \partial\Delta$ is a simple closed curve that either

(1) lies in a two-simplex of T,
(2) intersects three edges of T exactly once that meet at a vertex of Δ (see the left of Figure 3.5), or
(3) intersects four edges of T exactly once that are adjacent to an edge of Δ (see the right of Figure 3.5).

If C is not of this form, then there is a sub-arc b of an edge of Δ and a sub-arc a' of C such that $|\text{Int}(a') \cap T^1| \in \{0, 3\}$. The curve $a' \cup b$ bounds a bigon D' in $\partial\Delta$ whose interior is either disjoint from the one-skeleton or contains a single vertex of Δ; see Figure 3.6.

In either case, let a be an arc in $S \cap \Delta$ such that $a \cap \partial\Delta = \partial a'$, and such that a and a' are parallel. Then there is a disk D in Δ parallel to D' such that $\partial D \cap S = a$ and $\partial D \cap \partial\Delta = b$. (Note that the interior of D might intersect S.) Then a Whitney move (see Section 2.2) supported in $N(D)$ eliminates the intersection points ∂b from $S \cap T^1$, contradicting the minimality of $w(S)$.

Let $m(S)$ be the sum of $|S \cap \delta|$ over two-simplices δ of T, and isotope S such that $m(S)$ is minimal. Suppose that C is an innermost component of $S \cap \partial\Delta$ for a three-simplex Δ that lies in a two-simplex of Δ; that is, C is of type (1). Then C bounds a disk D in the complement of S. Since S is incompressible, this means that C also bounds a disk D' in S. Then $D \cup D'$ is a two-sphere that bounds a three-ball B in M, since M is irreducible. Hence, we can eliminate C from the intersection of S with two-simplices of T by isotoping S across B, decreasing $m(S)$. So we can assume that no component of $S \cap \partial\Delta$ is of the form (1).

Now suppose that there is a three-simplex Δ of T such that a component F of $S \cap \Delta$ is not a disk. Let C be a component of ∂F; this bounds a disk D' in

$\partial\Delta$. We choose F, C, and D' such that D' does not contain in its interior the boundary of a component of $S \cap \Delta$ that is not a disk. Then there is a disk $D \subset \Delta$ with $D \cap S = \partial D = C$. Since S is incompressible, we can again eliminate C by an isotopy of S, decreasing $m(S)$. Hence S is a normal surface when $m(S)$ is minimal. This concludes the proof of statement (i).

The proof of statement (ii) proceeds similarly, with the exception of the last two paragraphs. Suppose that C is an innermost component of $S \cap \partial\Delta$ of type (1). Let the components of S be $S_1, \dots, S_k$, and suppose that C lies in S_1. If we compress S_1 along D, we obtain the two-spheres S_1' and S_1''. Then at least one of $S' = S_1' \cup S_2 \cup \cdots \cup S_k$ and $S'' = S_1'' \cup S_2 \cup \cdots \cup S_k$ is independent. Indeed, suppose there were components B' and B'' of $M \setminus S'$ and $M \setminus S''$, respectively, that were punctured spheres. Since $S_1, \dots, S_k$ are independent, we must have $S_1' \subset \partial B'$ and $S_1'' \subset \partial B''$. If $S_1 \subset B'$ or $S_1 \subset B''$, then S would not be independent by the Schönflies theorem (Theorem 3.1). Otherwise, $B' \cap B'' = D$, so $B' \cup B''$ would be a punctured sphere component of $M \setminus S$, which is again impossible. Note that $w(S') < w(S)$ and $w(S'') < w(S)$. Hence, when $m(S)$ is minimal, there is no component C of $S \cap \partial\Delta$ that lies in a two-simplex. Similarly, we can arrange that $S \cap \Delta$ has only disk components for every three-simplex Δ of T. □

Definition 3.22 Let S and S' be disjoint embedded surfaces in the three-manifold M. Then we say that S and S' are *parallel* if $M \setminus (S \cup S')$ has a component C with $\partial C = S \cup S'$ and $\overline{C} \approx S \times I$ (in particular, $S \approx S'$).

Theorem 3.23 *For any closed, irreducible three-manifold M, there exists a constant $h(M) \in \mathbb{N}$, such that, for any incompressible surface S in M with $|S| \geq h(M)$, there are two components of S that are parallel.*

Proof Let T be a triangulation of M with t tetrahedra. Then we let

$$h(M) := 6t + 2b_2(M),$$

where $b_2(M) = \dim H_2(M; \mathbb{Z}_2)$. By statement (i) of Lemma 3.21, we can isotope S such that it becomes normal with respect to T.

If Δ is a three-simplex of T, a component of $\partial\Delta \setminus S$ is called *good* if it is an annulus, and it is *bad* otherwise. At most six components of $\partial\Delta \setminus S$ are bad. (The worst-case scenario is when we have all four types of triangles and one type of quad in Δ, in which case the four components containing the vertices and two components between triangles and quads are bad.) A component X of $M \setminus S$ is *good* if every component of $X \cap \partial\Delta$ is good for every three-simplex Δ of T, and it is *bad* otherwise. At most $6t$ components of $M \setminus S$ are bad. If

$|S| \geq 6t+2b_2(M)$, then $|M \setminus S| \geq 6t+1+b_2(M)$, so there are at least $1+b_2(M)$ good components.

A good component X is obtained by gluing regions homeomorphic to triangle times I and square times I along their edges times I. So, X is an I-bundle over a surface. Every non-trivial I-bundle contributes one $\mathbb{Z}_2$ direct summand to $H_2(M;\mathbb{Z}_2)$, so there is at least one trivial I-bundle X. The boundary components of X are parallel components of S. □

This finiteness result allows one to enumerate all normal surfaces by writing down so-called matching equations. The following result and corollary are due to Kneser [88]:

Theorem 3.24 *Let M be a closed triangulated three-manifold with t three-simplices. If M contains an independent set of k two-spheres, then*

$$k \leq 6t + 2b_2(M).$$

Proof The proof is analogous to the proof of Theorem 3.23, except now we invoke statement (ii) of Lemma 3.21. □

Corollary 3.25 *Every closed three-manifold can be expressed as a connected sum of finitely many prime three-manifolds.*

Uniqueness was shown 30 years later by Milnor [113].

Theorem 3.26 *Let M be a closed oriented three-manifold. If $M \approx M_1 \# \cdots \# M_k$ and $M \approx N_1 \# \cdots \# N_l$ with M_i, N_j prime and not S^3, then $k = l$, and after reindexing, M_i and N_i are orientation-preserving homeomorphic.*

Proof First, suppose that every two-sphere in M is separating. Then every M_i and N_j is irreducible. Let S be the union of the connected sum spheres in $M_1 \# \cdots \# M_k$. If Σ is a two-sphere separating N_1 and $N_2 \# \cdots \# N_l$, then we can isotope it such that it is transverse to S. Furthermore, we choose S to minimise $|S \cap \Sigma|$. We write $M_1^*, \ldots, M_k^*$ for the closures of the components of $M \setminus S$, where M_i^* is a punctured M_i. We define $N_1^*, \ldots, N_l^*$ analogously.

If $S \cap \Sigma \neq \emptyset$, then let C be a component of $S \cap \Sigma$ innermost in Σ, bounding a disk $D \subset \Sigma$ with $D \cap S = C = \partial D$; see the top of Figure 3.7. Suppose $D \subset M_i^*$. Since M_i is irreducible, one of the components of $M_i^* \setminus D$ has closure a punctured ball P. Suppose that C lies in a component S_j of ∂M_i^*, and let $D' = \overline{S_j \setminus P}$. Replace S_j by $D \cup D'$, pushed slightly to eliminate the intersection component C. This contradicts the choice of S to minimise the number of components of $S \cap \Sigma$.

So $S \cap \Sigma = \emptyset$. Suppose that $S \cap N_1^* \neq \emptyset$. Since N_1 is irreducible, some component of S is parallel to Σ and $M_i^* \subset N_1^*$ with $N_1^* \setminus M_i^* \approx S^2 \times I$ for some $i \in \{1, \ldots, k\}$. If $S \cap N_1^* = \emptyset$, then $N_1^* \subset M_i^*$ for some i. Since M_i is irreducible, $\overline{M_i^* \setminus N_1^*}$ is a punctured ball. In either case, $M_i \approx N_1$ and

$$M_1 \# \cdots \# M_{i-1} \# M_{i+1} \# \cdots \# M_k \approx N_2 \# \cdots \# N_l.$$

Hence, the result follows by induction.

Now suppose that F is a maximal collection of two-spheres in M such that $M \setminus F$ is connected, and let $r = |F|$. Furthermore, let S be as earlier. If $F \cap S \neq \emptyset$, then we choose a component C of $F \cap S$ innermost in S, and compress F along the disk that C bounds in S. This gives rise to collections of spheres F' and F'', one of which – say F' – does not separate; see the middle of Figure 3.7. Then we can isotope F' to eliminate C. Repeating this procedure, we obtain a sequence of non-separating collections of spheres $F = F^0, F^1, \ldots, F^n$ such that $F^n \cap S = \emptyset$. If $F_j^n \subset M_i^*$ for a component F_j^n of F^n, then $M_i \approx S^1 \times S^2$ by Exercise 3.17, since M_i is prime but not irreducible, as it contains the non-separating sphere F_j^n. Hence, after reindexing, $M_1 \approx \cdots \approx M_r \approx S^1 \times S^2$. Furthermore, $M \setminus N(F^n)$ is a $2r$-punctured $M_{r+1} \# \cdots \# M_k$. The same applies to the factorisation $N_1 \# \cdots \# N_l$.

We claim that $\overline{M \setminus N(F)} \approx \overline{M \setminus N(F^n)}$. It suffices to show that, if D is a disk in M with $D \cap F = \partial D$ such that ∂D bounds a disk $D' \subset F$ and $F' := (F \setminus D') \cup D$ does not separate M, then $\overline{M \setminus N(F)} \approx \overline{M \setminus N(F')}$; see the bottom of Figure 3.7. Indeed, since F was maximal, $D \times I$ separates $\overline{M \setminus N(F)}$ into two components, whose closures we denote by A and B. Then $\overline{M \setminus N(F)} \approx A \cup_D B$ and $\overline{M \setminus N(F')} \approx A \cup_{D'} B$. These are homeomorphic since both are obtained from $A \sqcup B$ by gluing disks in their boundaries.

We conclude that a $2r$-punctured $M_{r+1} \# \cdots \# M_k$ is homeomorphic to a $2r$-punctured $N_{r+1} \# \cdots \# N_l$, as they are both homeomorphic to $M \setminus N(F)$. Hence,

$$M_{r+1} \# \cdots \# M_k \approx N_{r+1} \# \cdots \# N_l,$$

and it does not contain a separating two-sphere, so the result now follows from the first part of the proof. $\square$

3.5 Heegaard Decompositions and Diagrams

A *genus g handlebody* is a three-ball with g oriented one-handles attached. Alternatively, it can be described as a regular neighbourhood of a wedge of g unknotted circles in $\mathbb{R}^3$. Every closed, connected, and oriented three-manifold can be obtained by gluing together two handlebodies of the same genus along their boundaries via an orientation-reversing diffeomorphism.

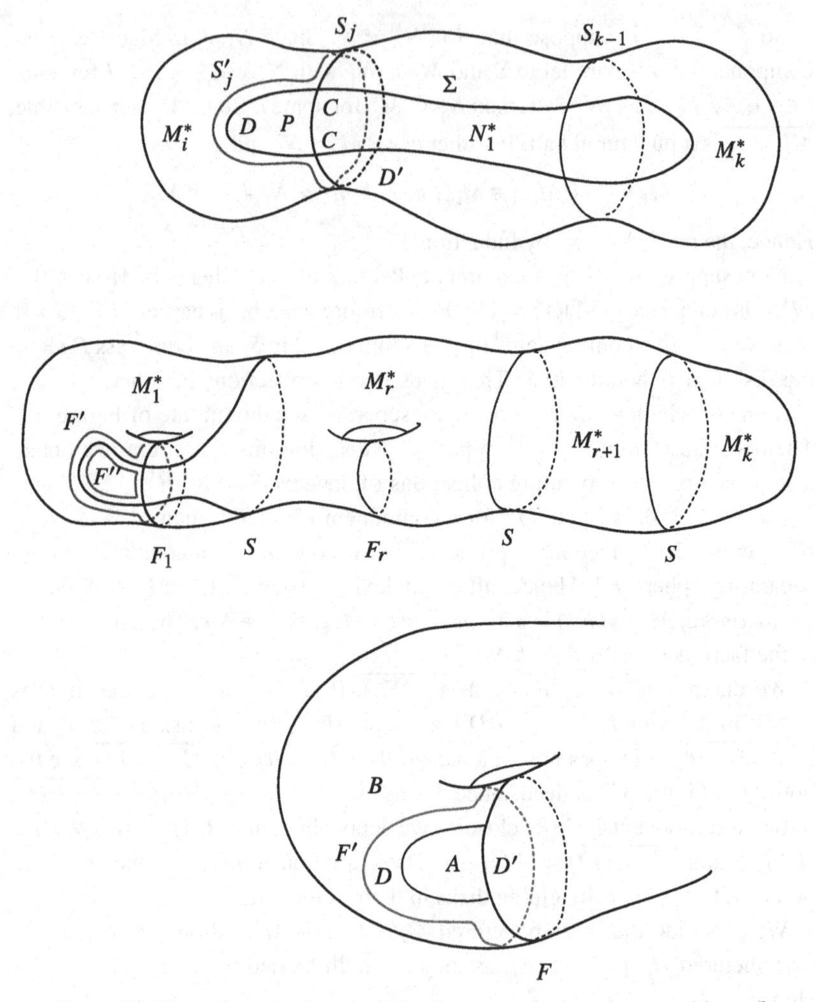

Figure 3.7 Illustrations for the proof of uniqueness of prime decompositions of three-manifolds.

Definition 3.27 Let M be a closed, connected, and oriented three-manifold. A *Heegaard decomposition* of M consists of a closed, connected, and oriented surface $\Sigma \subset M$ that separates M into the union of two handlebodies H_α and H_β, where Σ is oriented as the boundary of H_α. We call Σ a *Heegaard surface*. The *Heegaard genus* of M is the minimal genus of a Heegaard surface Σ.

Proposition 3.28 *Every closed, connected, and oriented three-manifold M admits a Heegaard decomposition.*

Proof Choose a *self-indexing* Morse function $f \colon M \to \mathbb{R}$; that is, if $c \in \mathrm{Crit}_i(f)$, then $f(c) = i$. Since M is connected, we can assume that f has a single index zero and a single index three critical point. Indeed, if there is more than one index zero critical point, we can cancel one of them against an index one critical point until we are left with just one.

Then f induces a handle decomposition of M with one zero-handle and one three-handle. The union of the zero-handle and the one-handles is one of the handlebodies, and the union of the three-handle and the two-handles is another handlebody. (The latter can be observed by considering $-f$.) The two handlebodies have the same genus since they share the same boundary, namely the surface $f^{-1}(3/2)$. □

Remark 3.29 An alternative proof often found in the literature uses the fact that every three-manifold admits a triangulation. Then a Heegaard decomposition can be obtained by taking a regular neighbourhood of the one-skeleton and its complement. The complement is a handlebody because it is a regular neighbourhood of the one-skeleton of the dual CW decomposition, whose zero-cells are the centres of the tetrahedra of the original triangulation, and two zero-cells are connected by a one-cell if and only if the corresponding tetrahedra share a face. It is a rather technical result that every smooth manifold admits a triangulation; see Munkres [124].

Given two genus g handlebodies, the diffeomorphism class of the three-manifold obtained by gluing them together along their boundaries via an orientation-reversing diffeomorphism depends only on the isotopy class of the gluing map. The group of isotopy classes of orientation-preserving automorphisms of a genus g surface Σ_g is called the *mapping class group* and is denoted by $\mathrm{MCG}(\Sigma_g)$. Due to Proposition 3.28, and since every handlebody admits an orientation-reversing symmetry, one can study three-manifolds via the mapping class group.

Exercise 3.30 Show that if M is a three-manifold, then $\pi_1(M)$ has a presentation with the same number of generators and relations. Show that $\mathbb{Z}^4$ is not the fundamental group of a closed three-manifold.

Since every orientation-reversing automorphism of S^2 is isotopic to a reflection, the only three-manifold of Heegaard genus zero is S^3. Heegaard genus one manifolds form an important class:

Definition 3.31 A *lens space* is a closed, connected, and oriented three-manifold of Heegaard genus one that is not $S^1 \times S^2$.

A Heegaard decomposition is not unique. For example, S^3 has a genus g Heegaard decomposition for every g: just consider the standard genus g surface in S^3. By a result of Waldhausen [174], every genus g Heegaard splitting of S^3 is isotopic to this.

Given three-manifolds M and M', together with Heegaard surfaces Σ and Σ', respectively, we can take their *connected sum*: Choose three-balls $B \subset M$ and $B' \subset M'$ such that both $B \cap \Sigma$ and $B' \cap \Sigma'$ are two-disks. Then, if we take the connected sum of M and M' along B and B', then $\Sigma \# \Sigma'$ is a Heegaard surface of $M \# M'$. Haken's lemma states that, if the three-manifold N contains an essential two-sphere and Σ is a Heegaard surface, then there is an essential two-sphere S such that $S \cap \Sigma$ is a single curve.

Definition 3.32 Let Σ be a Heegaard surface in M. Then a *stabilisation* of Σ is obtained by taking the connected sum with (S^3, T^2).

The following theorem is due to Reidemeister and Singer:

Theorem 3.33 *Let Σ_0 and Σ_1 be Heegaard surfaces in the closed, connected, and oriented three-manifold M. Then there is a Heegaard surface Σ that is isotopic to a stabilisation of both.*

Proof This is a simple application of Cerf theory. Choose self-indexing Morse functions f_0 and f_1 on M such that $f_i^{-1}(3/2) = \Sigma_i$ for $i \in \{0,1\}$. By Lemma 2.75, these can be connected by a generic one-parameter family of smooth functions f_t for $t \in I$ such that the Cerf graphic has a number of births, followed by crossings between critical points of the same index, and finally some deaths. Furthermore, whenever f_t is Morse, it is ordered. We can also arrange that $3/2$ is a regular level set of every f_t that is Morse. In particular, index one/two births and deaths occur on the $3/2$ level set.

Let us write $\Sigma_t := f_t^{-1}(3/2)$ for $t \in I$. Then index zero/one and one/two births or deaths do not change Σ_t, while index one/two births correspond to stabilisations, and index one/two deaths to destabilisations. If all births happen before $s \in (0,1)$ and all deaths happen after s, then $\Sigma := \Sigma_s$ is isotopic to stabilisations of Σ_0 and Σ_1. □

If one would also like to encode how the two handlebodies are glued together, one can record the belt circles of the one-handles of the two handlebodies:

Definition 3.34 An *abstract Heegaard diagram* is a triple $(\Sigma, \boldsymbol{\alpha}, \boldsymbol{\beta})$, where Σ is a closed, connected, and oriented, genus g surface, and $\boldsymbol{\alpha} = \alpha_1 \cup \cdots \cup \alpha_g$

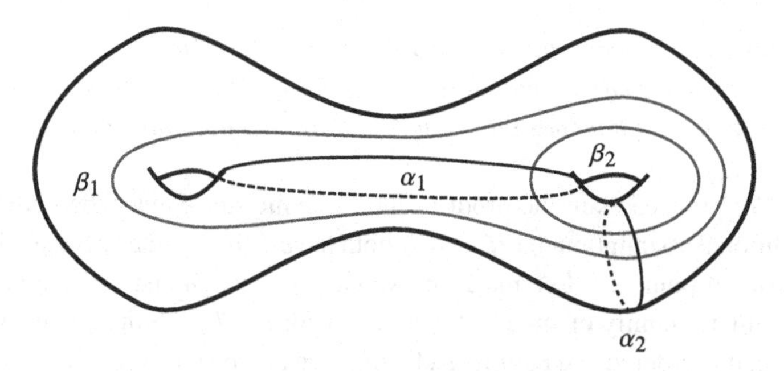

Figure 3.8 A Heegaard diagram.

and $\boldsymbol{\beta} = \beta_1 \cup \cdots \cup \beta_g$ are two g-component one-dimensional submanifolds of Σ whose components are linearly independent in $H_1(\Sigma)$; see Figure 3.8.

An *embedded Heegaard diagram* of the closed, connected, and oriented three-manifold M is an abstract Heegaard diagram $(\Sigma, \boldsymbol{\alpha}, \boldsymbol{\beta})$ such that $\Sigma \subset M$ is a Heegaard surface, each α_i bounds a disk in H_α, and each β_j bounds a disk in H_β.

For example, (T^2, μ, λ) is a genus 1 embedded Heegaard diagram of S^3, where μ is a meridian and λ is a longitude of T^2; see Remark 1.11. A *stabilisation* of a Heegaard diagram is obtained by taking the connected sum with this.

An abstract Heegaard diagram $\mathcal{H} = (\Sigma, \boldsymbol{\alpha}, \boldsymbol{\beta})$ defines a three-manifold as follows. Take $\Sigma \times I$, and attach three-dimensional two-handles along $\alpha \times \{0\}$ and $\beta \times \{1\}$ for every component α of $\boldsymbol{\alpha}$ and β of $\boldsymbol{\beta}$. Since the components of $\boldsymbol{\alpha}$ are linearly independent in $H_1(\Sigma)$ and there is g of them, the lower boundary component, which is obtained by surgering Σ along $\boldsymbol{\alpha}$, is a two-sphere. Similarly, the upper boundary component is also S^2. After attaching two three-balls to these two spheres, we obtain a closed, connected, and oriented three-manifold, which we call the three-manifold defined by $\mathcal{H}$. If $\mathcal{H}$ is an embedded Heegaard diagram of M, then the three-manifold defined by $\mathcal{H}$ is diffeomorphic to M relative to Σ.

If $\mathcal{H} = (\Sigma, \boldsymbol{\alpha}, \boldsymbol{\beta})$ is a Heegaard diagram of M, then we say that

$$\boldsymbol{\alpha}' = \boldsymbol{\alpha} \setminus \alpha_i \cup \alpha_i'$$

is obtained from $\boldsymbol{\alpha}$ by *sliding* α_i over α_j if $\alpha_i \cup \alpha_i' \cup \alpha_j$ is the boundary of a pair-of-pants (i.e., $S^2 \setminus P$ for $|P| = 3$) in Σ disjoint from the other curves in $\boldsymbol{\alpha}$. (This corresponds to sliding the three-dimensional two-handle corresponding to α_i over the handle corresponding to α_j in the three-manifold defined by the diagram.) We can define a handle slide among $\boldsymbol{\beta}$ analogously. Then we have the following strengthening of the Reidemeister–Singer theorem (Theorem 3.33):

Theorem 3.35 *Suppose that $\mathcal{H}$ and $\mathcal{H}'$ are embedded Heegaard diagrams of the closed, connected, oriented three-manifold M. Then they become isotopic after a sequence of handle slides, stabilisations, and destabilisations.*

Proof We proceed as in the proof of the Reidemeister–Singer theorem, with the additional assumption that f_0 and f_1 both have a single index zero and index three critical point. By Lemma 2.76, we can connect f_0 and f_1 by a generic one-parameter family of smooth functions f_t for $t \in I$ such that, whenever f_t is Morse, it is ordered and has precisely one index zero and three critical points. Furthermore, we can arrange that $f_t(\mathrm{Crit}_1(f_t)) < 3/2$ and $f_t(\mathrm{Crit}_2(f_t)) > 3/2$ for every $t \in I$.

Fix a Riemannian metric on M. Then $\Sigma_t := f_t^{-1}(3/2)$ is a Heegaard surface, and the α-curves correspond to the intersections of Σ_t with the unstable manifolds of index one critical points of f_t, and the β-curves to the intersections of Σ_t with the the stable manifolds of the index two critical points of f_t under the gradient flow of f_t. Index one/two births correspond to stabilisations, index one/two deaths to destabilisations, and gradient flow lines between index one (respectively two) critical points to α (respectively β) handle slides. □

Exercise 3.36 Determine the three-manifold defined by the Heegaard diagram in Figure 3.8.

3.6 Reidemeister Torsion

This section is based on Turaev [173, section 10]. We defined the torsion of an acyclic chain complex in Section 2.4. Reidemeister torsion of a finite CW complex builds on this construction. It was introduced by Reidemeister [153] in order to classify lens spaces up to homeomorphism.

Let X be a finite, connected CW complex. Then there is an induced CW structure on the universal cover $p\colon \tilde{X} \to X$. We orient the open cells of X and $\tilde{X}$ such that p is orientation-preserving on them. Let us write $\pi = \pi_1(X, x)$ for some basepoint $x \in X$. The covering action of π on $\tilde{X}$ induces an action of π on the chain groups $C_k(\tilde{X})$. This endows $C_k(\tilde{X})$ with a $\mathbb{Z}[\pi]$-module structure, and the boundary maps are $\mathbb{Z}[\pi]$-module homomorphisms.

Let $\{e_i^k\}$ be an ordering of the k-cells of X, and let $\tilde{e}_i^k \subset \tilde{X}$ be a lift of e_i^k. Then $\{\tilde{e}_i^k\}$ is a basis of $C_k(\tilde{X})$ over $\mathbb{Z}[\pi]$. Hence $C_*(\tilde{X})$ is a free and based chain complex over $\mathbb{Z}[\pi]$. However, it is not acyclic, as, for example, $H_0(C_*(\tilde{X}))$=$H_0(\tilde{X}; \mathbb{Z}) = \mathbb{Z}$, which is a torsion $\mathbb{Z}[\pi]$-module.

In order to obtain an acyclic chain complex, we proceed as follows. Let R be a commutative ring. Given a ring homomorphism $\phi\colon \mathbb{Z}[\pi] \to R$, consider the chain complex

$$C_*^\phi(X) := R \otimes_\phi C_*(\tilde{X}).$$

Then $C_k^\phi(X) = \oplus_i R\tilde{e}_i^k$, and the matrix of the boundary map

$$\partial_k^\phi : C_k^\phi(X) \to C_{k-1}^\phi(X)$$

is obtained by applying ϕ to the matrix of $\partial_k : C_k(\tilde{X}) \to C_{k-1}(\tilde{X})$. The *twisted homology* of X is defined as $H_*^\phi(X) := H(C_*^\phi(X))$.

Definition 3.37 Suppose that $C_*^\phi(X)$ is acyclic; that is, its homology is zero. Then the *Reidemeister torsion* of X is defined to be

$$\tau_\phi(X) := \tau(C_*^\phi(X)) \in R^*/\pm\phi(\pi);$$

see Definition 2.10.

Exercise 3.38 Show that $\tau_\phi(X) \in R^*/\pm\phi(\pi)$ is independent of the orientations and ordering of the cells, and the choice of their lifts.

The following result is due to Chapman [22]:

Theorem 3.39 *Reidemeister torsion is invariant under homeomorphisms of CW complexes.*

3.7 The Classification of Lens Spaces

Recall that a lens space is a closed, connected, and oriented three-manifold of Heegaard genus one other than $S^1 \times S^2$. In other words, a manifold that can be obtained by gluing two solid tori along their boundaries via an orientation-reversing automorphism, with the exception of S^3 and $S^1 \times S^2$. The gluing map is described by an automorphism of T^2. Since

$$\pi_0(\mathrm{Diff}^+(T^2)) \cong \mathrm{SL}(2, \mathbb{Z}),$$

every orientation-reversing automorphism of T^2 is isotopic to a transformation given by a matrix

$$\begin{pmatrix} q & r \\ p & s \end{pmatrix}$$

with determinant -1, so p and q are relatively prime. If m is the meridian and l is the longitude, then m is mapped to $qm + pl$. This already determines the resulting three-manifold $L(p, q)$, since every automorphism of T^2 that preserves m extends to $S^1 \times D^2$.

Alternatively, we can also describe $L(p, q)$ as the quotient of $S^3 \subset \mathbb{C}^2$ by the action of the cyclic group C_p given by

$$(z, w) \mapsto (\zeta z, \zeta^q w)$$

for a pth root of unity $\zeta \in C_p$. This is a free action, hence the universal cover of $L(p,q)$ is S^3, and

$$\pi_1(L(p,q)) \cong C_p.$$

We orient $L(p,q)$ such that this covering map is orientation-preserving. Alternatively, one can compute the fundamental group by applying the Seifert–van Kampen theorem to the first description of lens spaces. A *canonical generator* of $\pi_1(L(p,q))$ is one that is represented by the core of one of the solid tori in the genus one Heegaard decomposition. The following discussion is based on the book of Turaev [173, section 10].

Theorem 3.40 *Let $L = L(p,q)$ for $p > 2$ be a lens space, $T \in \pi_1(L)$ a canonical generator, and $r \in \mathbb{Z}_p$ such that $rq \equiv 1 \mod p$. If $\phi\colon \mathbb{Z}[\pi_1(L)] \to \mathbb{F}$ is a ring homomorphism to a field $\mathbb{F}$ such that $H_*^\phi(L) = 0$ and $t := \phi(T) \neq 1$, then the Reidemeister torsion*

$$\tau_\phi(L) = (t-1)^{-1}(t^r - 1)^{-1} \in \mathbb{F}^*/\pm\{t^j\}_{j\in\mathbb{Z}_p}.$$

Note that $t^r \neq 1$ since $(t^r)^q = t \neq 1$.

Proof To compute the torsion, we construct a CW decomposition of L, coming from a C_p-equivariant CW decomposition of $S^3 \subset \mathbb{C}^2$. Let $\zeta := \mathrm{e}^{2\pi\mathrm{i}/p} \in S^1$ and consider the arcs

$$I_j := \{\mathrm{e}^{2\pi(j+t)\mathrm{i}/p} : t \in I\} \subset S^1$$

for $j \in \mathbb{Z}_p$. The closed balls

$$\begin{aligned}
e_j^0 &:= \{(\zeta^j, 0)\},\\
e_j^1 &:= I_j \times \{0\},\\
e_j^2 &:= \{(z_1, t\zeta^j)\colon t \in I, |z_1|^2 = 1 - t^2\},\\
e_j^3 &:= \{(z_1, z_2) \in S^3 : z_2 \in I_j \cdot I \subset \mathbb{C}\}
\end{aligned}$$

in $\mathbb{C}^2$ for $j \in \mathbb{Z}_p$ give a C_p-equivariant CW decomposition of S^3; see Figure 3.9. Note that e_j^k is a k-ball. Furthermore, $\partial e_j^2 = S^1 \times \{0\}$, $\partial e_j^3 = e_j^2 \cup e_{j+1}^2$, and $S^1 \times \{0\}$ is the one-skeleton. We orient the zero-cells $e_0^0, \ldots, e_{p-1}^0$ positively. The one-cells are oriented consistently with S^1; that is, $\partial e_j^1 = e_{j+1}^0 - e_j^0$, where $e_p^0 := e_0^0$. The two-cells are oriented such that $\partial e_j^2 = e_0^1 + \cdots + e_{p-1}^1$, and finally e_j^3 is oriented such that $\partial e_j^3 = e_{j+1}^2 - e_j^2$, where $e_p^2 = e_0^2$. This determines the boundary map on $C_*(S^3)$.

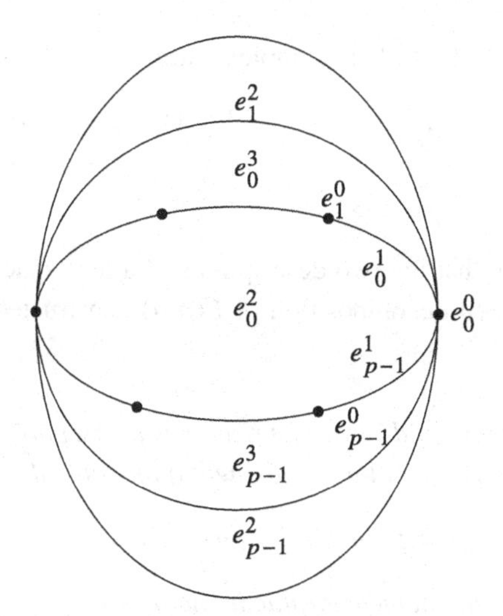

Figure 3.9 A C_p-equivariant cell decomposition of S^3.

Recall that $\zeta \in C_p$ acts on S^3 via $\zeta \cdot (z_1, z_2) = (\zeta z_1, \zeta^q z_2)$. Hence, the action of ζ on the cells is given by $\zeta \cdot e_j^0 = e_{j+1}^0$, $\zeta \cdot e_j^1 = e_{j+1}^1$, $\zeta \cdot e_j^2 = e_{j+q}^2$, and $\zeta \cdot e_j^3 = e_{j+q}^3$. So the chain complex $C_*(S^3)$ is of the form

$$0 \to \mathbb{Z}[C_p]e_0^3 \to \mathbb{Z}[C_p]e_0^2 \to \mathbb{Z}[C_p]e_0^1 \to \mathbb{Z}[C_p]e_0^0 \to 0$$

with boundary map

$$\partial e_0^1 = (\zeta - 1)e_0^0,$$
$$\partial e_0^2 = \left(\sum_{j=0}^{p-1} \zeta^j\right) e_0^1,$$
$$\partial e_0^3 = (\zeta^r - 1)e_0^2.$$

Under the canonical isomorphism $\pi_1(L) \cong C_p$, the element $T \in \pi_1(L)$ corresponds to $\zeta \in C_p$. By assumption, $t = \phi(T) \neq 1 \in \mathbb{F}^*$. Since

$$t^p = \phi(T^p) = \phi(1) = 1,$$

we have $\sum_{j=0}^{p-1} t^j = \frac{t^p - 1}{t-1} = 0$. Hence, the chain complex

$$C_*^\phi(L) = \mathbb{F} \otimes_{\mathbb{Z}[C_p]} C_*(S^3)$$

is obtained by concatenating the chain complexes

$$\xrightarrow{0} \mathbb{F}e_0^1 \xrightarrow{t-1} \mathbb{F}e_0^0 \xrightarrow{0} \quad \text{and} \quad \xrightarrow{0} \mathbb{F}e_0^3 \xrightarrow{t^r-1} \mathbb{F}e_0^2 \xrightarrow{0}.$$

Since $t \neq 1$ and $t^r \neq 1$, these chain complexes are acyclic. Hence, by definition,

$$\tau_\phi(L) = \tau(C_*^\phi(L)) = (t-1)^{-1}(t^r-1)^{-1} \in \mathbb{F}^*/\pm\{t^j\}_{j\in\mathbb{Z}_p},$$

as claimed. □

Exercise 3.41 Show that the two descriptions of a lens space are equivalent. (Hint: consider the cell decomposition of $L(p,q)$ constructed in the proof of Theorem 3.40.)

Theorem 3.42 *Let (p,q) and (p',q') be relatively prime pairs of integers. Then $L(p,q)$ and $L(p',q')$ are homotopy equivalent if and only if*

$$p = p' \text{ and } qq' \equiv \pm n^2 \mod p$$

for some $n \in \mathbb{Z}$. They are homeomorphic if and only if

$$p = p' \text{ and } q' \equiv \pm q^{\pm 1} \mod p.$$

Proof As $\pi_1(L(p,q)) \cong C_p$, we have $p = p'$ if $L(p,q)$ and $L(p',q')$ are homotopy equivalent. We only prove the statement about homeomorphism.

Note that $L(p,q) = L(p,q+p)$, since $\zeta^p = 1$ for every $\zeta \in C_p$. Consider the map $c(z,w) = (z,\overline{w})$ for $(z,w) \in \mathbb{C}^2$. Since $\overline{\zeta^q w} = \zeta^{-q}\overline{w}$, the restriction $c|_{S^3}$ induces a diffeomorphism from $L(p,q)$ to $L(p,-q)$. This is orientation-reversing. It follows that each lens space is diffeomorphic to a lens space $L(p,q)$ with $1 \leq q < p/2$.

We now show that $L(p,q)$ and $L(p,q^{-1})$ are diffeomorphic, though the diffeomorphism is orientation-reversing. This is induced by the automorphism of $\mathbb{C}^2$ given by $(z,w) \mapsto (w,z)$, where the image is quotiented by the action $(w,z) \mapsto (\zeta^q w, \zeta z)$. If we choose the generator $\zeta^{q^{-1}}$ of C_p, then we get the same quotient, but the action becomes $(w,z) \mapsto (\zeta w, \zeta^{q^{-1}} z)$. The quotient by this action is $L(p,q^{-1})$, by definition. The generator $\zeta \in C_p$ defines a canonical generator of $\pi_1(L(p,q))$. The preceding diffeomorphism does not preserve the canonical generators.

By Theorem 3.40, if $\phi\colon \pi_1(L(p,q)) \to \mathbb{C}$ maps the canonical generator to ζ and $r = q^{-1}$, then

$$\tau_\phi(L) = (\zeta-1)^{-1}(\zeta^r-1)^{-1} \in \mathbb{C}^*/\pm\{\zeta^j\}_{j\in\mathbb{Z}_p}.$$

So, if $L(p,q)$ and $L(p,q')$ are homeomorphic via a homeomorphism that preserves the canonical generators of π_1 and $r' = (q')^{-1}$, then

$$(\zeta-1)^{-1}(\zeta^r-1)^{-1} = \pm\zeta^j(\zeta-1)^{-1}(\zeta^{r'}-1)^{-1}$$

for some $j \in \mathbb{Z}$. After dividing both sides by $(\zeta - 1)^{-1}$, multiplying the resulting equation with its conjugate, and taking inverses, we obtain that

$$(\zeta^r - 1)(\zeta^{-r} - 1) = (\zeta^{r'} - 1)(\zeta^{-r'} - 1),$$

from which $\mathrm{Re}(\zeta^r) = \mathrm{Re}(\zeta^{r'})$. It follows that $r' = \pm r$.

The action $(z, w) \mapsto (\zeta^k z, \zeta^{kq} w)$ also has quotient $L(p, q)$, but replaces the distinguished generator with its kth power for k prime to p. It follows from the preceding that $L(p, q)$ and $L(p, q')$ are homeomorphic if and only if $(\zeta^k, \zeta^{kq'}) = (\zeta, \zeta^{\pm q})$ or $(\zeta^k, \zeta^{kq'}) = (\zeta^{\pm q}, \zeta)$ for some $k \in \mathbb{Z}$; that is, $q' \equiv \pm q^{\pm 1} \mod p$. □

It follows from Theorem 3.42 that the lens spaces $L(7, 1)$ and $L(7, 2)$ are homotopy equivalent but not homeomorphic. This was the key property that Milnor used to show that the manifolds $L(7, 1) \times S^4$ and $L(7, 2) \times S^4$ are h-cobordant but not diffeomorphic; see Section 2.3.

4
Knots and Links

Knots and links play a fundamental role in low-dimensional topology. Their exteriors provide an important class of three-manifolds. One can obtain all three-manifolds by using surgery on links in S^3 or by taking branched covers of S^3 along links. Furthermore, all four-manifolds can be obtained from traces of surgeries on links in connected sums of copies of $S^1 \times S^2$ (after adding three-handles and a four-handle in a unique way); see Section 6.1 on Kirby calculus for more information. Invariants of three-manifolds are often easier to understand for links, and some of them are in fact defined as extensions of link invariants.

Many knot invariants are derived from the Seifert form, which is defined using a compact oriented surface the knot bounds in the three-sphere. These include the Alexander polynomial, the signature, and the Arf invariant. We introduce several important classes of knots, such as alternating knots, torus knots, satellite knots, hyperbolic knots, and two-bridge knots.

We discuss the knot group, which is the fundamental group of the knot complement, and how it can be computed using the Wirtinger presentation and via Fox calculus. We then prove a characterisation of fibred knots in terms of the knot group due to Stallings, and another one using sutured manifolds due to Gabai.

The Jones polynomial is a more modern but topologically less well-understood knot invariant. It is defined in terms of knot diagrams, which are projections of the knot into the plane. Two diagrams represent the same knot if and only if they can be connected by a sequence of Reidemeister moves. To show that the Jones polynomial is well defined, one checks that it is unchanged by these moves. The volume conjecture provides a potential link between the Jones polynomial and the volume of a hyperbolic knot.

Two knots are concordant if they cobound an annulus in $I \times S^3$. Concordance classes of knots form a group that is still quite mysterious and is the focus of active research. The signature and the Arf invariant are classical concordance invariants, and we will encounter some more modern ones when we study Heegaard Floer homology.

We conclude this chapter by giving an overview of two-knots, which are two-spheres embedded in the four-sphere. This gives an introduction to some techniques in four-manifold topology.

4.1 Knots and Links

Definition 4.1 A smooth (respectively topological) *k-component n-link* in an $(n+2)$-manifold M is a smooth (respectively topologically locally flat) embedding

$$L\colon \bigcup_{i=1}^{k} S^n \times \{i\} \hookrightarrow M.$$

The links L_0 and L_1 are *equivalent* if they are ambient isotopic; that is, there is an isotopy $\Phi\colon M \times I \to M$ such that $\Phi_0 = \mathrm{Id}_M$ and $\Phi_1 \circ L_0 = L_1$. An *n-knot* is a one-component n-link. We simply write 'knot' for a one-knot and 'link' for a one-link.

In this book, we will mostly focus on one-links and two-links. From now on, all links will be assumed to be one-links, unless otherwise stated.

One could also define piecewise linear (PL) knots and links, whose study is amenable to combinatorial techniques. However, we will take the differential-topological point of view. The two are equivalent for one-links.

The image $\mathrm{Im}(L)$ is a smooth one-dimensional submanifold of M that is oriented such that the map L is orientation-preserving. Given an orientation of $\mathrm{Im}(L)$, the parametrisation L is uniquely determined up to isotopy as $\mathrm{Diff}^+(S^1)$ is connected. Hence, we will often think of a link as an oriented smooth one-dimensional submanifold of M. We will denote by $\overline{L}$ the link L with its orientation reversed, called the *reverse* of L.

By work of Cerf [20], any orientation-preserving automorphism of S^3 is isotopic to the identity. So, two links in S^3 are ambient isotopic if and only if they are orientation-preserving diffeomorphic. We denote by $-L$ the *mirror* of the link L, obtained by reflecting L in a plane.

We can represent a link in $\mathbb{R}^3$ by considering its projection onto the xy-plane $\mathbb{R}^2$. If this projection is an immersion with only transverse double point singularities, and at each double point we record which strand is higher with respect to the z-coordinate, we obtain a link diagram. This allows one to reconstruct the link up to equivalence.

Definition 4.2 A *link diagram* is an immersed, oriented one-manifold in $\mathbb{R}^2$ or S^2 that has only transverse double point singularities, and at each double point, a designation of one of the two strands as the overstrand.

Every link has a diagram:

Proposition 4.3 *Let L be a link in $\mathbb{R}^3$. Then there is an isotopy $\{\psi_t : t \in I\}$ of $\mathbb{R}^3$ such that $\psi_0 = Id_{\mathbb{R}^3}$ and orthogonal projection of $\psi_1(L)$ onto $\mathbb{R}^2$ is an immersion with only transverse double point singularities.*

Proof This is similar to the proof of Theorem 1.43. First, note that it suffices to find a direction $v \in S^2$ such that, after slightly perturbing L by an isotopy, the projection of L to $v^\perp$ is an immersion with only transverse double points. Indeed, if $\{\phi_t : t \in I\}$ is a family of rotations such that $\phi_1(v) = \partial/\partial x_3$, then the projection of $\phi_1(L)$ to $\mathbb{R}^2$ has the required properties.

For $v \in S^2$, the projection of L onto $v^\perp$ is an immersion whenever v is not tangent to L. The unit tangent bundle of L is a one-manifold. Hence, its image in S^2 is a measure zero subset A_1 by Sard's theorem (Theorem 1.42).

Consider the map

$$f : (L \times L \times L) \setminus \Delta_L \to S^2 \times S^2,$$
$$(x, y, z) \mapsto \left(\frac{y - x}{|y - x|}, \frac{z - y}{|z - y|} \right)$$

where Δ_L is the diagonal of $L \times L \times L$ defined as

$$\Delta_L := \{(x, y, z) \in L \times L \times L : x = y \text{ or } y = z \text{ or } z = x\}.$$

By perturbing L slightly, we can make f transverse to the diagonal of $S^2 \times S^2$, defined as

$$\Delta_S := \{(x, y) \in S^2 \times S^2 : x = y\}.$$

Then $\mathrm{Im}(f) \cap \Delta_S$ is a manifold of dimension

$$\dim(L \times L \times L) + \dim(\Delta_S) - \dim(S^2 \times S^2) = 3 + 2 - 4 = 1.$$

We denote its projection to the first factor of $S^2 \times S^2$ by A_2. Then A_2 is also measure zero in S^2 by Sard's theorem. If $v \in S^2 \setminus A_2$, then each point of the projection of L to $v^\perp$ has at most two pre-images.

Finally, consider the map

$$g : (L \times L) \setminus \Delta \to S^2,$$
$$(x, y) \mapsto \frac{y - x}{|y - x|}$$

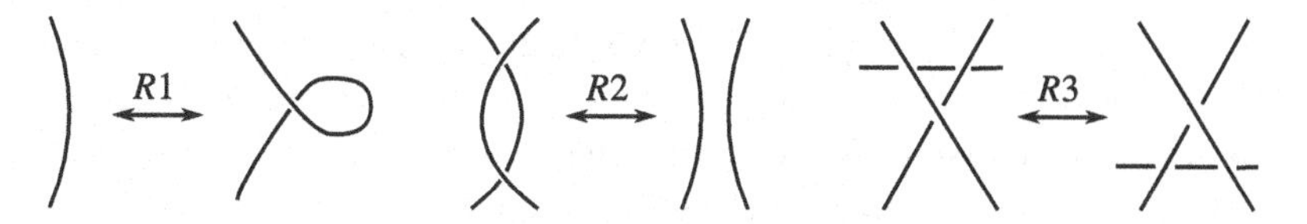

Figure 4.1 The three Reidemeister moves.

where Δ is the diagonal of $L \times L$. Let A_3 be the set of critical values of g, which is measure zero in S^2 by Sard's theorem. Let $v \in S^2 \setminus A_3$, and write $p_v : L \to v^\perp$ for the orthogonal projection. Suppose that $x, y \in L$ are distinct points such that $p_v(x) = p_v(y)$; that is, $g(x,y) = \pm v$. Then the unit tangents of L at x and y project to linearly independent vectors in $v^\perp = T_v S^2$ since the differential

$$dg : T_{(x,y)}((L \times L) \setminus \Delta) \to T_v S^2$$

is non-singular as v and $-v$ are regular values of g.

In conclusion, if $v \in S^2 \setminus (A_1 \cup A_2 \cup A_3)$, then the image of the orthogonal projection $p_v : L \to v^\perp$ is an immersion with only self-transverse double points. □

4.2 Reidemeister Moves

Given two link diagrams, they represent equivalent links if and only if they can be connected by a sequence of Reidemeister moves and planar isotopies. We describe these next. For an illustration, see Figure 4.1.

We say that two diagrams are related by a *planar isotopy* if there is an isotopy of $\mathbb{R}^2$ taking one diagram to the other.

Let $\mathcal{D}$ be a diagram of the link L. We can perform the *first Reidemeister move R1* if there is a component C of $\mathbb{R}^2 \setminus \mathcal{D}$ that contains a single crossing in its boundary. The resulting diagram is obtained by smoothing $\overline{\mathcal{D} \setminus \partial C}$, and has one fewer crossings than $\mathcal{D}$. We will also refer to the reverse of this move as an R1 move.

Now suppose that $\mathbb{R}^2 \setminus \mathcal{D}$ contains a component C that has two crossings along its boundary, and such that $\overline{C} \approx D^2$. We further assume that, as we traverse one of the two strands along ∂R, both crossings are either overcrossings, or they are both undercrossings. Then the *second Reidemeister move R2* applied to the component C removes a neighbourhood of C from $\mathcal{D}$ and reconnects the resulting two pairs of endpoints in $N(C)$ without intersection points, as in the middle of Figure 4.1. Again, we also call the reverse of this move an R2 move.

Finally, suppose that $\mathbb{R}^2 \setminus \mathcal{D}$ has a component C with $\overline{C} \approx D^2$ whose boundary contains three crossings, and one of the three strands along ∂C – call it s – has two undercrossings; see the right-hand side of Figure 4.1. Let c be the crossing along ∂C opposite s, and let C' be the component of $\mathbb{R}^2 \setminus \mathcal{D}$ that meets C at c. Then we can remove a small extension of s from $\mathcal{D}$ and reconnect its endpoints across C'. This is called the *third Reidemeister move R3.*

The following result is due to Reidemeister [152]:

Theorem 4.4 *Let $\mathcal{D}_0$ and $\mathcal{D}_1$ be link diagrams. Then they represent equivalent links if and only if they can be connected by a sequence of planar isotopies and Reidemeister moves R1–R3.*

Sketch of proof We denote by $\pi\colon \mathbb{R}^3 \to \mathbb{R}^2$ the projection. Let L_i be a link with diagram $\mathcal{D}_i$ for $i \in \{0, 1\}$. Suppose there is an isotopy ϕ_t for $t \in I$ of $\mathbb{R}^3$ such that $\phi_0 = \mathrm{Id}_{S^3}$ and $\phi_1(L_0) = L_1$. We write $L_t := \phi_t(L)$. If the isotopy ϕ_t is generic, then $\pi(L_t)$ is an immersion with only transverse double points for every $t \in I$ outside a finite set $S \subset (0, 1)$. Furthermore, for $s \in S$, the projection $\pi(L_s)$ has exactly one of the following singularities:

(i) a cusp, corresponding to a point $p \in L_t$ where $T_p L_t$ is parallel to $\partial/\partial z$,
(ii) a self-tangency modelled on the graphs of x^2 and $-x^2$,
(iii) a triple point where any two branches of the curve are transverse.

Indeed, these singularities all have codimension one by the proof of Proposition 4.3. We write $\mathcal{D}_t$ for the diagram of L_t for $t \in I \setminus S$. If t and t' lie in the same component of $I \setminus S$, then the diagrams $\mathcal{D}_t$ and $\mathcal{D}_{t'}$ are related by a planar isotopy by the isotopy extension theorem (Theorem 1.36). Furthermore, for $s \in S$ and ε sufficiently small, $\mathcal{D}_{s-\varepsilon}$ and $\mathcal{D}_{s+\varepsilon}$ are related by a Reidemeister move R1 in case (i), move R2 in case (ii), and move R3 in case (iii).

Conversely, every planar isotopy and Reidemeister move can be lifted to an isotopy of the link. Furthermore, if two links have the same diagram, then they are isotopic via linearly interpolating between the z-coordinate functions. □

A corollary of the preceding theorem is that any quantity that is defined for link diagrams and is invariant under Reidemeister moves gives rise to a link invariant. The Jones polynomial, which we will encounter in Section 4.8, is a particularly interesting example of this construction.

Definition 4.5 Let $\mathcal{D}$ be a link diagram. We say that a crossing of $\mathcal{D}$ is *positive* if, as we traverse the overstrand following the orientation, the understrand crosses from right to left, and is *negative* otherwise.

The *writhe* $w(\mathcal{D})$ of the diagram $\mathcal{D}$ is the number of positive crossings minus the number of negative crossings.

The *winding number* of the knot diagram $\mathcal{D}$ is the number of rotations the unit tangent vector of $\mathcal{D}$ makes as we traverse the knot.

Recall from Exercise 2.70 that two immersed curves in the plane are regularly homotopic if and only if they have the same winding number. Writhe and winding number are *not* knot invariants, but are only changed by move R1. Trace [172] showed that two diagrams of the same knot are related only using moves R2 and R3 if and only if they have the same writhe and winding number.

The writhe is closely related to the notion of linking. By Alexander duality, $H_1(S^3 \setminus K) \cong \mathbb{Z}$ for any knot K. Given disjoint knots K and K', their *linking number* $\mathrm{lk}(K, K')$ is the integer corresponding to $[K'] \in H_1(S^3 \setminus K)$. Given a diagram for the link $K \cup K'$, this can be computed by taking the number of positive crossings minus the number of negative crossings between K and K', divided by two.

Definition 4.6 Given knots K_1 and K_2, we can form their *connected sum* $K_1 \# K_2$, as follows: Choose diagrams $\mathcal{D}_1$ and $\mathcal{D}_2$, respectively, such that $\mathcal{D}_1 \subset \mathbb{R}^2_+$ and $\mathcal{D}_2 \subset \mathbb{R}^2_-$. Let $r\colon I \times I \hookrightarrow \mathbb{R}^2$ be an embedding such that

$$r(I \times I) \cap \mathcal{D}_1 = r(\{0\} \times I) \text{ and } r(I \times I) \cap \mathcal{D}_2 = r(\{1\} \times I),$$

and the orientation along $\partial r(I \times I)$ is coherent with the orientations of $\mathcal{D}_1$ and $\mathcal{D}_2$. Then $K_1 \# K_2$ has diagram the symmetric difference $(\mathcal{D}_1 \cup \mathcal{D}_2) \triangle (\partial r(I \times I))$.

Definition 4.7 The crossing number $c(\mathcal{D})$ of a diagram $\mathcal{D}$ of the link L is the number of its crossings. We define the *crossing number* $c(L)$ of the link L to be the minimum of $c(\mathcal{D})$ over all diagrams of L.

Knots and links are usually tabulated according to their crossing number. For example, the Rolfsen table of knots [155] tabulates all knots up to ten crossings. The *unknot* is represented by the standard circle S^1 in S^3 and is the only knot with crossing number zero. There is no knot of crossing number one or two. Up to mirroring, there is just one knot of crossing number three: the *trefoil* (see Figure 4.5), and one crossing number four knot: the *figure eight*. Even though the crossing number is easy to define, it is unknown whether it is additive under connected sums.

Coward and Lackenby [26] provided an upper bound on the number of Reidemeister moves required to pass between two diagrams of the same link.

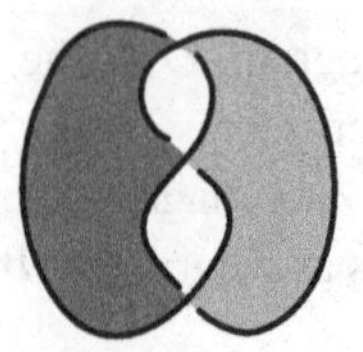

Figure 4.2 A Seifert surface for the trefoil knot.

However, the upper bound is a tower of exponentials in the sum of crossing numbers of the two diagrams.

Lackenby [93] also showed that one can always reduce any diagram $\mathcal{D}$ of the unknot to the trivial diagram in at most $(236c(\mathcal{D}))^{11}$ Reidemeister moves. There are so-called *hard unknot diagrams*, where one has to first increase the crossing number in any sequence of Reidemeister moves to the trivial diagram.

4.3 Seifert Surfaces

Surfaces bounding knots in the three-sphere play an important role in defining various knot invariants.

Definition 4.8 Let L be a link in the three-manifold Y. A *Seifert surface* for L is a compact, oriented surface S embedded in Y without closed components such that $\partial S = L$.

See Figure 4.2 for a Seifert surface for the trefoil knot.

Proposition 4.9 *Every link in S^3 admits a Seifert surface.*

Proof Let $\mathcal{D}$ be a diagram of L. We resolve each crossing according to the orientation of $\mathcal{D}$, obtaining an embedded, oriented one-manifold C in $\mathbb{R}^2$ whose components are called *Seifert circles*. We choose a collection S_0 of disjoint disks in $\mathbb{R}^3$ with boundary C. For each crossing in $\mathcal{D}$, we add a half-twisted band to S_0, obtaining the desired Seifert surface S. □

We say that the diagram $\mathcal{D}$ is *simple* if it has no nested Seifert circles. The preceding proof only works for links in S^3. More generally, we have the following result:

Proposition 4.10 *Let L be a null-homologous oriented $(n-2)$-manifold in the n-manifold Y, and $a \in H_{n-1}(Y, L)$ a homology class such that $\partial a = [L] \in$*

$H_{n-2}(L)$, where $\partial\colon H_{n-1}(Y,L) \to H_{n-2}(L)$ is the boundary map in the long exact sequence of the pair (Y,L). Then there is a compact, connected, oriented $(n-1)$-manifold embedded in Y with boundary L representing a.

Proof Let $E := Y \setminus N(L)$ be the exterior of L. Since L is null-homologous in Y, there is a trivialisation $\partial E \approx L \times S^1$ such that $L \times \{1\}$ is null-homologous in E. By excision,

$$H_{n-1}(Y,L) \cong H_{n-1}(E,\partial E).$$

Let $\alpha \in H^1(E)$ be the dual of the class in $H_{n-1}(E,\partial E)$ corresponding to $a \in H_{n-1}(Y,L)$. Then $\alpha(\mu) = 1$ for any meridian μ of L since $\partial a = [L]$.

As the Eilenberg–MacLane space $K(\mathbb{Z},1) \cong S^1$, the map

$$i\colon [E,S^1] \to H^1(E)$$

given by $i(f) := f^*(1)$ for $[f] \in [E,S^1]$ and $1 \in H^1(S^1)$ is an isomorphism. Let $f_\alpha\colon E \to S^1$ be a map such that $i(f_\alpha) = \alpha$ and $f_\alpha|_{\partial E}\colon L \times S^1 \to S^1$ is projection onto the second factor. We can perturb f_α such that it is smooth, and such that $1 \in S^1$ is a regular value. After taking the connected sum of the components of $f^{-1}(\{1\})$, it extends in $N(L)$ to a compact, connected, oriented manifold with boundary L representing a. □

Seifert surfaces are not unique. For example, one can always choose a three-disk $D \subset Y$ such that $D \cap S \approx D^2 \sqcup D^2$, and replace $D \cap S$ with an unknotted annulus. This operation is called a *stabilisation*, and its reverse a *destabilisation*.

Proposition 4.11 *Let L be a null-homologous link in the three-manifold Y and $a \in H_2(Y,L)$ a homology class. Then any two Seifert surfaces of L representing the class a can be connected by a sequence of isotopies, stabilisations, and destabilisations.*

Proof Let S_0 and S_1 be Seifert surfaces of L representing the class a, and write E for the link exterior. Choose maps f_0, $f_1\colon E \to S^1$ such that 1 is a common regular value of both, $f_0^{-1}(\{1\}) = S_0$ and $f_1^{-1}(\{1\}) = S_1$, and which are projections onto the S^1-factor along $\partial N(L) \approx L \times S^1$.

Let $\{f_t : t \in I\}$ be a generic one-parameter family of smooth functions connecting f_0 and f_1 relative to ∂E. Then f_t is a circle-valued Morse function, except for finitely many values of t, when f_t has a single birth-death critical point p, and generically $f(p) \neq 1$; see Section 2.9.

Hence, the pre-image $f_t^{-1}(\{1\})$ only changes when there is a non-degenerate critical point p of f_t with $f_t(p) = 1$. If 1 is a regular value of f_t, we write S_t

for the connected sum of the components of $f_t^{-1}(\{1\})$. When the value of an index 0 or 3 critical point passes through 1, the pre-image $f_t^{-1}(\{1\})$ changes by the birth or death of a small two-sphere that bounds a three-ball, hence $S_{t-\varepsilon}$ and $S_{t+\varepsilon}$ are isotopic. When the value of an index 1 or 2 critical point passes through 1, the level set $f_t^{-1}(\{1\})$ changes by adding or removing a tube. Hence, the surfaces $S_{t-\varepsilon}$ and $S_{t+\varepsilon}$ are related by an isotopy when the tube connects distinct components, or a stabilisation or destabilisation otherwise. □

Remark 4.12 While a knot in S^3 has no unique Seifert surface, the normal framing given by a Seifert surface is well-defined, and is called the *Seifert framing*. This follows by applying the half lives, half dies lemma to the knot exterior. The *blackboard framing* of a knot diagram is given by the normal in the plane of the diagram. The difference between the Seifert framing and the blackboard framing is the writhe of the diagram.

Definition 4.13 Let K be a knot in S^3. Then its *Seifert genus* $g(K)$ is the minimum of $g(S)$, where S is a Seifert surface of K and $g(S)$ is the genus of S.

The Seifert genus detects the unknot, in the following sense:

Lemma 4.14 *Let K be a knot in S^3. Then $g(K) = 0$ if and only if K is the unknot U.*

Proof Clearly, $g(U) = 0$. Conversely, suppose that $g(K) = 0$. Then there is an embedding $e\colon D^2 \hookrightarrow S^3$ such that $e|_{S^1} = K$. Then $e_t := e|_{(1-t)\cdot S^1}$ for $t \in [0, 1-\varepsilon]$ provides an isotopy of K to the curve $e_{1-\varepsilon}$. If ε is small enough, then we can linearly isotope $e_{1-\varepsilon}$ to the tangent plane $e_*(T_0D^2)$ and obtain the unknot. □

The following celebrated result is due to Haken [54]:

Theorem 4.15 *There is an algorithm to decide whether a knot in S^3 is trivial.*

Proof The only knot K in S^3 with exterior $S^1 \times D^2$ is the unknot. Indeed, $\{1\} \times \partial D^2$ has to be a longitude of K for homological reasons, so $g(K) = 0$ and $K = U$ by Lemma 4.14. A knot exterior is a Haken three-manifold, so we can use Theorem 3.18 to decide whether it is homeomorphic to $S^1 \times D^2$. □

Proposition 4.16 *Let K_1 and K_2 be knots in S^3. Then*

$$g(K_1 \# K_2) = g(K_1) + g(K_2).$$

Proof By considering the boundary connected sum of Seifert surfaces of K_1 and K_2, we obtain that $g(K_1 \# K_2) \leq g(K_1) + g(K_2)$.

We now show that $g(K_1 \# K_2) \geq g(K_1) + g(K_2)$. Let S be a Seifert surface for $K_1 \# K_2$. Make S transverse to the connected sum sphere C. Then $C \cap S$ is a one-manifold with a single arc component, and a collection of circle components. Choose an innermost circle component γ of $C \cap S$, which bounds a disk $D \subset C$ such that $D \cap S = \gamma$. If we compress S along D, we obtain a surface S' such that $\chi(S') = \chi(S) + 2$. If S' has an S^2 component, we remove it and obtain a surface with Euler characteristic $\chi(S)$.

Repeating this process, we obtain a surface S'' with boundary K such that $S'' \cap C$ is a single arc and $\chi(S'') \geq \chi(S)$. Furthermore, S'' has no S^2 component. Hence, removing the closed components of S'', we obtain a Seifert surface S''' of K such that $\chi(S''') \geq \chi(S)$, and so $g(S''') \leq g(S)$. As $S''' \cap C$ consists of a single component, S''' is a boundary connected sum of Seifert surfaces of K_1 and K_2. Hence $g(S) \geq g(S''') \geq g(K_1) + g(K_2)$, as claimed. □

Definition 4.17 We say that a knot $K \neq U$ is *prime* if $K = K_1 \# K_2$ implies that $K_1 = U$ or $K_2 = U$.

Corollary 4.18 *Every non-trivial knot can be written as a connected sum of prime knots.*

Proof We prove the claim by induction on $g(K)$. If K is not prime, then it can be written as $K = K_1 \# K_2$, where $K_1 \neq U$ and $K_2 \neq U$. Furthermore, $g(K_1) < g(K)$ and $g(K_2) < g(K)$ by Proposition 4.16 and Lemma 4.14. We can now write K_1 and K_2 as a connected sum of prime knots by the inductive hypothesis. □

Schubert [157] proved the uniqueness of prime decompositions of knots:

Theorem 4.19 *Any two prime decompositions of a knot are related by a permutation of the summands.*

4.4 The Seifert Form

We now introduce the Seifert form, which is the source of a number of powerful knot invariants.

Definition 4.20 Let S be a Seifert surface of a knot in S^3. We define the *Seifert form*

$$\langle\, , \rangle_S \colon H_1(S) \times H_1(S) \to \mathbb{Z}$$

as follows. For $a, b \in H_1(S)$, choose one-cycles $\alpha, \beta \subset S$ representing them. We write β^+ for the one-cycle in $S^3 \setminus S$ obtained by pushing β off S along the positive normal of S. Then

$$\langle a, b\rangle_S := \mathrm{lk}(\alpha, \beta^+).$$

If $a_1, \ldots, a_{2g}$ is a basis of the free $\mathbb{Z}$-module $H_1(S)$, then the corresponding *Seifert matrix* V has (i, j)-th entry $\langle a_i, a_j\rangle_S$ for $i, j \in \{1, \ldots, 2g\}$.

The Seifert form is bilinear, but not necessarily symmetric. Given a Seifert matrix V of K, we call $V + V^T$ the corresponding *symmetrised Seifert matrix*.

Definition 4.21 The *Alexander polynomial* of a knot K in S^3 is defined as

$$\Delta_K(t) := \det(V - tV^T),$$

where V is a Seifert matrix for K.

Proposition 4.22 *The Alexander polynomial is well defined up to multiplication by t^k for $k \in \mathbb{Z}$.*

Proof We first show that $\det(V - tV^T)$ is independent of the choice of basis $a_1, \ldots, a_{2g}$. Let $b_1, \ldots, b_{2g}$ be another basis of $H_1(S)$. Then the Seifert matrix V_1 obtained from $b_1, \ldots, b_{2g}$ is of the form $M^T VM$ for an invertible integral matrix M. Then $\det(M) = \pm 1$, as it is a unit in $\mathbb{Z}$. Hence

$$\begin{aligned}\det(V_1 - tV_1^T) = \det(M^T VM - tM^T V^T M) = \\ \det(M^T)\det(V - tV^T)\det(M) = \\ \det(V - tV^T).\end{aligned}$$

By Proposition 4.11, it suffices to prove that $\det(V - tV^T)$ is unchanged by a stabilisation of S. Suppose that S' is a stabilisation of S. If V is the Seifert matrix of S, then

$$V' = \begin{pmatrix} V & * & 0 \\ * & * & 1 \\ 0 & 0 & 0 \end{pmatrix} \text{ or } \begin{pmatrix} V & * & 0 \\ * & * & 0 \\ 0 & 1 & 0 \end{pmatrix}$$

is a Seifert matrix of S', where we extend a basis of $H_1(S)$ by adding a longitude and a meridian of the stabilisation tube. After a change of basis, this becomes

$$\begin{pmatrix} V & * & 0 \\ 0 & 0 & 1 \\ 0 & 0 & 0 \end{pmatrix} \text{ or } \begin{pmatrix} V & 0 & 0 \\ * & 0 & 0 \\ 0 & 1 & 0 \end{pmatrix}.$$

Hence, we have $\det(V' - t(V')^T) = t\det(V - tV^T)$, and the result follows. □

The Alexander polynomial satisfies the symmetry relation $\Delta_K(t) \doteq \Delta_K(t^{-1})$, where '$\doteq$' denotes equality up to multiplication by t^k for some $k \in \mathbb{Z}$. The *Conway normalisation* of the Alexander polynomial is the unique Laurent polynomial satisfying

$$\Delta_K(t) = \Delta_K(t^{-1}).$$

It is immediate from the definition that

$$\deg \Delta_K(t) \leq g(K).$$

Since $V - V^T$ agrees with the matrix of the intersection form of S in the basis $a_1, \dots, a_{2g}$, which is unimodular, we have

$$\Delta_K(1) = \det(V - V^T) = \pm 1. \tag{4.1}$$

Furthermore, the *determinant* of K is defined as

$$\det(K) := \det(V + V^T) = \Delta_K(-1).$$

We now introduce the signature of a knot. This, as we shall see, gives a lower bound on the genera of orientable surfaces the knot bounds in the four-ball.

Definition 4.23 The *signature* $\sigma(K)$ of a knot K is defined as the signature of the symmetrised Seifert matrix $V + V^T$ (i.e., the number of positive eigenvalues minus the number of negative eigenvalues).

Proposition 4.24 *The signature $\sigma(K)$ of a knot K is well defined.*

Proof For a given basis of $H_1(S)$, the symmetrised Seifert matrix is the matrix of the symmetric bilinear form

$$\langle a, b \rangle := \langle a, b \rangle_S + \langle b, a \rangle_S.$$

The signature is then the dimension of a maximal positive definite subspace of $H_1(S)$ minus the dimension of a maximal negative definite subspace of $H_1(S)$. This shows that the signature is independent of the choice of basis.

Invariance under stabilisation of the Seifert surface follows from the observation that, up to change of basis, $V + V^T$ changes by taking the direct sum with the matrix

$$\begin{pmatrix} 0 & 1 \\ 1 & 0 \end{pmatrix},$$

which has signature zero. See the proof of Proposition 4.22. □

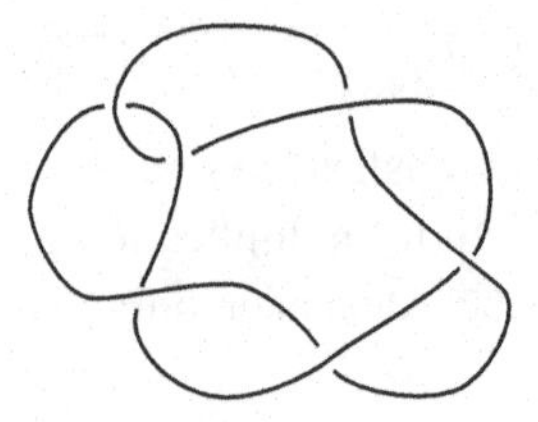

Figure 4.3 The twist knot 6_1 is the first non-trivial slice knot.

It immediately follows from the definition that the signature is additive under connected sum. We have the following analogues of Seifert surfaces and the Seifert genus in dimension four:

Definition 4.25 Let K be a knot in S^3. A *slice surface* for K is a compact, connected, and oriented surface F embedded in D^4 such that $\partial F = K$. The *four-ball genus* $g_4(K)$ of K is the minimal genus of a slice surface for K.

Clearly, $g_4(K) \leq g(K)$. A knot is called *slice* if $g_4(K) = 0$. There are non-trivial slice knots. The smallest one is the twist knot 6_1 in the Rolfsen table; see Figure 4.3.

Exercise 4.26 Show that the knot 6_1 is slice by attaching a single band, resulting in a two-component unlink.

Lemma 4.27 *Let K and K' be disjoint knots in S^3 with Seifert surfaces S and S' and slice surfaces F and F', respectively. Then*

$$lk(K, K') = S \cdot K' = S' \cdot K = F \cdot F',$$

where $\cdot$ denotes the algebraic intersection number.

Proof Since S is Poincaré–Lefschetz dual to the meridian of K in the exterior of K, we have $\text{lk}(K, K') = S \cdot K'$. Similarly, $\text{lk}(K, K') = S' \cdot K$.

View D^4 as S^4_+, and push S' into S^4_-. Then $F \cup -S$ and $F' \cup -S'$ are closed, oriented surfaces in S^4 that are null-homologous, hence

$$0 = (F \cup -S) \cdot (F' \cup -S') = F \cdot F' + S \cdot S'.$$

As $S \cap S' = S \cap K'$, and since the intersection signs are opposite, the result follows from the previous paragraph. □

Proposition 4.28 *Let K be a knot in S^3. Then*

$$\frac{|\sigma(K)|}{2} \leq g_4(K).$$

Proof Let $g = g_4(K)$, and let F be a genus g slice surface for K. Choose an arbitrary Seifert surface S for K. Then $S \cup F$ is a closed, oriented, null-homologous surface in D^4, and hence bounds a Seifert three-manifold M according to Proposition 4.10. Let

$$U = \ker(H_1(\partial M) \to H_1(M)).$$

By the half lives, half dies lemma (Lemma 3.12),

$$\mathrm{rk}(U) = \frac{\mathrm{rk}(H_1(\partial M))}{2} = g(S) + g(F).$$

Note that $H_1(\partial M) \cong H_1(S) \oplus H_1(F)$, and write $U_S := U \cap H_1(S)$. Then

$$\mathrm{rk}(U_S) \geq \mathrm{rk}(U) - \mathrm{rk}(H_1(F)) = g(S) - g(F). \tag{4.2}$$

We claim that $\langle a, b\rangle_S = 0$ for $a, b \in U_S$. Indeed, a and b are null-homologous in M, and let A and B be two-chains in M with $\partial A = a$ and $\partial B = b$. We write B^+ for the two-chain in D^4 obtained by pushing B off M in the direction of the positive normal of M. Then $\partial B^+ = b^+$ and $A \cap B^+ = \emptyset$, hence

$$\langle a, b\rangle_S = \mathrm{lk}(a, b^+) = A \cdot B^+ = 0,$$

where the second equality follows from Lemma 4.27.

On the other hand, if V is a Seifert matrix associated to S, then

$$\det(V + V^T) \equiv \det(V - V^T) \equiv 1 \mod 2$$

by Equation (4.1), so $V + V^T$ is non-degenerate. Hence, if $P \leq H_1(S)$ and $N \leq H_1(S)$ are maximal positive and negative definite subspaces of the Seifert form, respectively, then $P \oplus N = H_1(S)$. So, if we write $p = \mathrm{rk}(P)$ and $n = \mathrm{rk}(N)$, then $p + n = 2g(S)$.

As $P \cap U_S = \{0\}$ and $N \cap U_S = \{0\}$, we have

$$\max(p, n) \leq 2g(S) - \mathrm{rk}(U_S) \leq g(S) + g(F),$$

where the last inequality follows from Equation (4.2). Hence

$$\sigma(K) = p - n = (p + n) - 2n \geq 2g(S) - 2(g(S) + g(F)) = -2g(F).$$

Similarly, $-\sigma(K) \geq -2g(F)$, and the result follows. □

As we shall see, *isotropic* subspaces of the Seifert from – subspaces on which the Seifert form vanishes, such as U_S in the above proof – play an important role when studying knot concordance.

The signature of a knot can also be computed from an unoriented spanning surface for the knot due to the work of Gordon and Litherland [50].

Definition 4.29 Let S be a compact, connected, embedded surface in S^3. The *Gordon–Litherland pairing*

$$\langle\, , \rangle_S \colon H_1(S) \times H_1(S) \to \mathbb{Z}$$

is defined as follows. Let a, $b \in H_1(S)$, represented by oriented multicurves α, $\beta \subset S$. Consider the unit normal bundle $p_S \colon UN(S) \to S$ of S in S^3. Then

$$\langle \alpha, \beta \rangle_S := \mathrm{lk}(\alpha, p_S^{-1}(\beta)).$$

The Gordon–Litherland pairing is symmetric and bilinear. It agrees with the symmetrised Seifert form when S is orientable. Let $\{b_1, \ldots, b_n\}$ be a basis of $H_1(S)$. Then the *Goeritz matrix* G_S is an $n \times n$ symmetric matrix with (i, j)-th entry $\langle b_i, b_j \rangle_S$ for $i, j \in \{1, \ldots, n\}$. Furthermore, the *normal Euler number* $e(S)$ of S is defined to be $-\mathrm{lk}(K, K')$, where K' is the framing of K given by S. Gordon and Litherland proved the following:

Theorem 4.30 *Let S be an unoriented surface bounding a knot K in S^3. Then*

$$\sigma(K) = \sigma(G_S) + \frac{e(S)}{2},$$

where $\sigma(G_S)$ is the signature of the Goeritz matrix.

Definition 4.31 The *Arf invariant* $\mathrm{Arf}(K)$ of a knot K is the Arf invariant of the quadratic form

$$q(a) := \mathrm{lk}(a, a^+) \mod 2$$

for some Seifert surface S of K and $a \in H_1(S; \mathbb{F}_2)$; see Definition 2.31. Concretely, if $a_1, b_1, \ldots, a_g, b_g$ is a symplectic basis of $H_1(S; \mathbb{F}_2)$, then

$$\mathrm{Arf}(K) = \sum_{i=1}^{g} \mathrm{lk}(a_i, a_i^+)\mathrm{lk}(b_i, b_i^+) \mod 2.$$

We checked independence of the Arf invariant from the choice of symplectic basis in Proposition 2.32. Invariance under stabilisation of the Seifert surface follows from the observation that we can extend a symplectic basis of S to a symplectic basis of the stabilised surface by adding a pair of basis elements a_{g+1} and b_{g+1} such that $\mathrm{lk}(a_{g+1}, a_{g+1}^+) = 0$, where a_{g+1} is the meridian of the stabilisation tube.

Proposition 4.32 *If the knot K is slice, then $\mathrm{Arf}(K) = 0$.*

Levine–Tristram signatures generalise the knot signature:

Definition 4.33 Let K be a knot in S^3, and V a Seifert matrix. Then the *Levine–Tristram signature* is a function $\sigma_K \colon S^1 \to \mathbb{Z}$ defined as the signature of the matrix

$$(1 - \omega)V + (1 - \overline{\omega})V^T$$

for $\omega \in S^1$.

As $(1 - \omega)V + (1 - \overline{\omega})V^T = (1 - \omega)(V - \overline{\omega}V^T)$ is non-singular when $\overline{\omega}$ is not a root of the Alexander polynomial Δ_K, the signature function σ_K is continuous, and hence constant away from the roots of $(1 - t)\Delta_K$. Note that $\sigma(K) = \sigma_K(-1)$.

4.5 Important Classes of Knots

In this section, we present several important classes of knots and links: alternating knots, torus knots, satellite knots, hyperbolic links, and two-bridge links.

4.5.1 Alternating Knots

Definition 4.34 The knot K is *alternating* if it admits a diagram $\mathcal{D}$ such that, as we travel along $\mathcal{D}$, we alternatingly encounter undercrossings and overcrossings.

Alternating knots have a number of nice properties. A large percentage of small-crossing knots are alternating, but their proportion tends to zero as the crossing number goes to infinity.

Definition 4.35 We say that a crossing c of a knot diagram $\mathcal{D}$ is *reducible* or *nugatory* if there is a circle $C \subset \mathbb{R}^2$ such that $C \cap \mathcal{D} = \{c\}$. The diagram $\mathcal{D}$ is called *reduced* if it does not have a reducible crossing.

We can always remove such a crossing by applying $\pm\pi$-rotation about a line in $\mathbb{R}^2$ through the crossing to the part of $\mathcal{D}$ in the circle C. Consequently, if a diagram realises the crossing number of a knot, it has to be reduced. A related operation on knot diagrams is called a flype.

Definition 4.36 Let $\mathcal{D}$ be a knot diagram, and suppose that $C \subset \mathbb{R}^2$ is a circle that intersects $\mathcal{D}$ in a crossing c and two additional points that are not

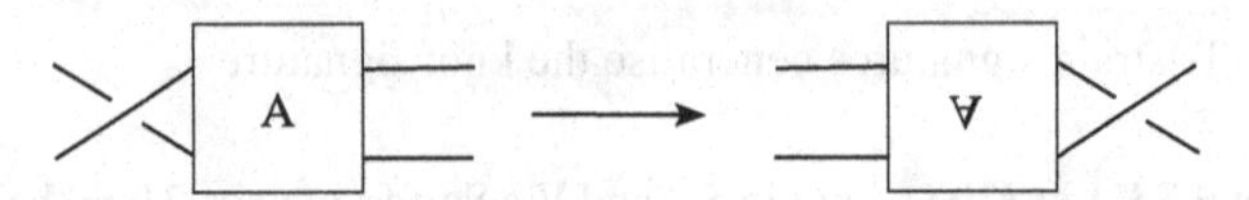

Figure 4.4 A flype consists of rotating the part of the knot in the box by angle π around the x-axis.

crossings. A *flype* on $\mathcal{D}$ is the operation of rotating the portion of $\mathcal{D}$ inside C by angle $\pm\pi$ about a line in $\mathbb{R}^2$ passing through c such that we remove the crossing c, but create a new crossing between the other two strands intersecting C; see Figure 4.4.

The following results were conjectured by Tait. The first two statements were shown by Kauffman [77], Murasugi [127], and Thistlethwaite [167] using the Jones polynomial, and the third by Menasco and Thistlethwaite [110].

Theorem 4.37 *Let $\mathcal{D}$ and $\mathcal{D}'$ be reduced diagrams of an alternating knot K. Then the following hold:*

(i) $c(\mathcal{D}) = c(\mathcal{D}')$,
(ii) $w(\mathcal{D}) = w(\mathcal{D}')$, *and*
(iii) *we can obtain $\mathcal{D}'$ from $\mathcal{D}$ using flypes.*

Statement (i) means that the crossing number of an alternating knot is realised by any reduced diagram. The reduced diagram is unique up to flypes according to statement (iii), known as the Tait flyping conjecture. Since a flype preserves both the crossing number and the writhe, statement (iii) implies statements (i) and (ii).

Theorem 4.38 *If we apply Seifert's algorithm to a reduced diagram of an alternating knot, we obtain a minimal genus Seifert surface.*

Definition 4.34 does not give an intrinsic characterisation of alternating knots, as it depends on diagrams. The following result of Greene [53] gives a more intrinsic description and also implies Tait's conjectures (i) and (ii):

Given an alternating diagram $\mathcal{D}$, we can colour the components of $\mathbb{R}^2 \setminus \mathcal{D}$ black and white in a chequerboard fashion, giving rise to two (not necessarily orientable) surfaces bounding the knot. One of these surfaces has positive definite, and the other has negative definite Gordon–Litherland pairing. Greene's characterisation of alternating knots is the converse of this statement:

Theorem 4.39 *A knot K in S^3 is alternating if and only if it bounds surfaces with both positive and negative definite Gordon–Litherland pairings.*

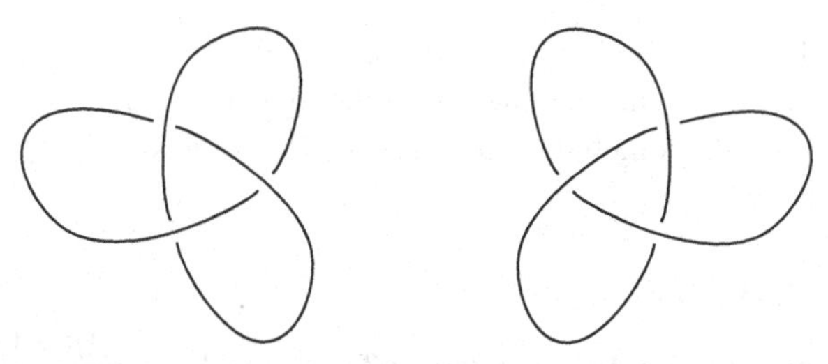

Figure 4.5 The drawing on the left shows the left-handed trefoil $T_{2,-3}$, and the one on the right its mirror, the right-handed trefoil $T_{2,3}$.

Given such surfaces, K admits an alternating diagram where the chequerboard surfaces are isotopic to the two given surfaces.

As a corollary, Greene re-proved the following converse of Theorem 4.38 for special alternating knots, due to Banks and Hirasawa–Sakuma [64]:

Corollary 4.40 *A Seifert surface for a special alternating knot K has minimal genus if and only if it is obtained by applying Seifert's algorithm to a special alternating diagram of K.*

4.5.2 Torus Knots

Definition 4.41 Let p and q be coprime integers. The (p, q)*-torus knot* $T_{p,q}$ is given by the curve on the standard torus T^2 in $\mathbb{R}^3$ that winds around the longitude p times and the meridian q times.

In other words, $T_{p,q}$ is the projection of the line $\langle(p, q)\rangle \subset \mathbb{R}^2$ under the covering map $\mathbb{R}^2 \to \mathbb{R}^2/\mathbb{Z}^2 \cong T^2$ that sends $(1,0)$ to the longitude and $(0, 1)$ to the meridian. When p and q are not coprime and n is their highest common factor, we obtain an n-component link, where each component is $T_{(p/n,q/n)}$. For example, $T_{2,2}$ is the Hopf link.

When $p = 1$ or $q = 1$, the torus knot $T_{p,q}$ is the unknot. The first non-trivial torus knot is $T_{2,3}$, which is the trefoil knot, denoted by 3_1 in the Rolfsen table; see Figure 4.5.

Torus knots are far from being alternating when p and q are large. They are particularly important as they arise from singularities of complex plane curves:

Exercise 4.42 The complex polynomial $f(w, z) = w^p + z^q$ has a critical point at the origin of $\mathbb{C}^2$. Let the plane curve $V_f \subset \mathbb{C}^2$ be the zero-set of f. Show that, for $\varepsilon > 0$ sufficiently small, the link of the singular point of V_f at the origin, defined as $V_f \cap S^3_\varepsilon$, is the torus knot $T_{p,q}$.

The torus knots $T_{p,q}$ and $T_{q,p}$ are equivalent. Indeed, there is an orientation-preserving automorphism of S^3 that swaps the solid tori on the two sides of T^2, and interchanges the longitude and meridian. The torus knots $T_{p,q}$ and $T_{p,-q}$ are mirror images.

Proposition 4.43 *Let p, $q > 0$. Then*

$$c(T_{p,q}) = \min\{(p-1)q, (q-1)p\} \text{ and } g(T_{p,q}) = g_4(T_{p,q}) = \frac{(p-1)(q-1)}{2}.$$

The standard diagram of $T_{p,q}$, where p parallel strands twist around a circle q times has $(p-1)q$ crossings, since at each twist an outermost strand passes over $p-1$ other strands. Similarly, the standard diagram of $T_{q,p}$ has $(q-1)p$ crossings. This implies that

$$c(T_{p,q}) \le \min\{(p-1)q, (q-1)p\}.$$

A sharp lower bound on the crossing number of torus knots was given by Murasugi [128] using the HOMFLY polynomial that we will introduce in Section 4.8.

The preceding formula for the four-ball genus of torus knots was conjectured by Milnor and can be shown using the knot invariant $\tau(K)$ defined using Heegaard Floer homology, which we will discuss in Section 5.3.

Exercise 4.44 Compute the signature of the right-handed trefoil using Seifert matrices, and also using the formula of Gordon and Litherland from the chequerboard surfaces.

4.5.3 Satellite Knots

Definition 4.45 Let K' be a knot in the solid torus $S^1 \times D^2$ that does not lie in a three-ball and is not isotopic to $S^1 \times \{0\}$. Furthermore, let K be a knot in S^3, called the *companion*, and $n \in \mathbb{Z}$ a framing coefficient. Choose a diffeomorphism $f \colon S^1 \times D^2 \to N(K)$ representing the framing n. Then the *n-twisted satellite of K with pattern K'* is $f(K')$, which we denote by $K(K', n)$.

Example 4.46 If there is a point $p \in S^1$ such that $|(\{p\} \times D^2) \cap K'| = 1$, then $K(K', n) = K \# K'$ for any n, where we view K' as a knot in S^3 by identifying $S^1 \times D^2$ with the standard solid torus in S^3.

Example 4.47 If $K' = T_{p,q}$, then $K(K', 0)$ is called the *(p, q)-cable* of K. The $(n, 0)$-cable of K is also known as the n-cable of K. (Note that $T_{n,0}$ is an n-component unlink.)

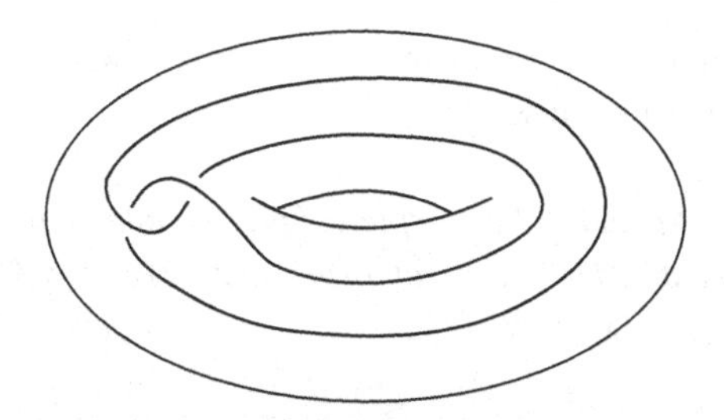

Figure 4.6 The Whitehead knot in the solid torus.

Example 4.48 If W is the Whitehead knot in $S^1 \times D^2$ shown in Figure 4.6, then $K(W, n)$ is called the *n-twisted Whitehead double* of K.

The *algebraic winding number* of an oriented knot in $S^1 \times D^2$ is its signed intersection number with $\{1\} \times D^2$ (or, equivalently, its homology class in $H_1(S^1 \times D^2) \cong \mathbb{Z}$), while its *geometric winding number* is the minimal number of intersection points with $\{1\} \times D^2$ counted without signs. The algebraic winding number of W is zero, while its geometric winding number is two.

Exercise 4.49 If K is a Whitehead double, it has a genus one Seifert surface. Use this to show that $\Delta_K(t) = 1$.

Satellite knots are characterised by the property that their exterior E contains a torus T not parallel to ∂E which is incompressible; that is, there is no homotopically non-trivial simple closed curve on T that bounds an embedded disk in E.

4.5.4 Hyperbolic Links

Definition 4.50 A link L in S^3 is *hyperbolic* if its complement $S^3 \setminus L$ admits a complete hyperbolic metric; that is, a Riemannian metric of constant sectional curvature -1.

The following is a special case of Thurston's hyperbolisation theorem [170]:

Theorem 4.51 *Every knot in the three-sphere is either hyperbolic, a torus knot, or a satellite, and these classes are mutually exclusive.*

In other words, we can build all knots from hyperbolic knots and torus knots using satellite operations. By the Mostow–Prasad rigidity theorem [123][146], which we now state, the hyperbolic structure is unique up to isometry, and so any geometric quantity one can assign to the hyperbolic structure on the knot complement is a knot invariant:

Theorem 4.52 *Let M and N be complete, finite-volume hyperbolic manifolds of dimensions at least 3. Then any isomorphism $\pi_1(M) \to \pi_1(N)$ is induced by a unique isometry from M to N.*

Consider the *Poincaré half-space model*

$$\mathbb{H}^n := \{ (x_1, \ldots, x_n) \in \mathbb{R}^n : x_n > 0 \}$$

of hyperbolic n-space with the metric

$$\frac{\mathrm{d}x_1^2 + \cdots + \mathrm{d}x_n^2}{x_n^2}.$$

Every complete, connected hyperbolic n-manifold has universal cover $\mathbb{H}^n$, hence can be written as $\mathbb{H}^n/\Gamma$, where Γ is a torsion-free discrete group of isometries of $\mathbb{H}^n$. When $n = 3$, this isometry group is $\mathrm{PSL}(2, \mathbb{C})$. Hence hyperbolic three-manifolds can also be studied using group theory.

If K is a hyperbolic knot in S^3, then we say that the subset $N \subset S^3 \setminus K$ is a maximal *cusp* if it is a maximal open subset isometric to

$$\{ (x_1, x_2, x_3) \in \mathbb{H}^3 : x_3 \geq 1 \}/G$$

for $G \leq \mathrm{PSL}(2, \mathbb{C})$ such that $G \cong \mathbb{Z} \oplus \mathbb{Z}$. Then $\partial N \approx \mathbb{C}/\Lambda$ is a Euclidean torus for some lattice Λ. We can arrange that the longitude of K lifts to $[0, \lambda]$ for some $\lambda \in \mathbb{R}_{\geq 0}$, which is called the *longitudinal translation* of K. Furthermore, we arrange that the meridian lifts to $[0, \mu]$ for some $\mu \in \mathbb{C}$ with $\mathrm{Im}(\mu) > 0$. This μ is called the *meridional translation* of K. It is known that $1 \leq |\mu| \leq 6$.

The most fundamental hyperbolic knot invariant is the *hyperbolic volume* $\mathrm{vol}(K)$, which is the volume of $S^3 \setminus K$. Other invariants describe the shape of a maximal cusp, such as the cusp volume, the lengths of the longitude and meridian, and the longitudinal and meridional translations, defined earlier.

The hyperbolic structure on a knot complement can often be found using the computer software SnapPy [27], which can also compute many of the associated invariants.

The *natural slope* of a hyperbolic knot $K \subset S^3$ is

$$\mathrm{slope}(K) := \mathrm{Re}(\lambda/\mu).$$

This measures the shape of the cusp torus $\mathbb{C}/\Lambda$. We define the *injectivity radius* of a hyperbolic knot K in S^3 as

$$\mathrm{inj}(K) := \inf\{ \mathrm{inj}_x(S^3 \setminus K) : x \in (S^3 \setminus K) \setminus N \},$$

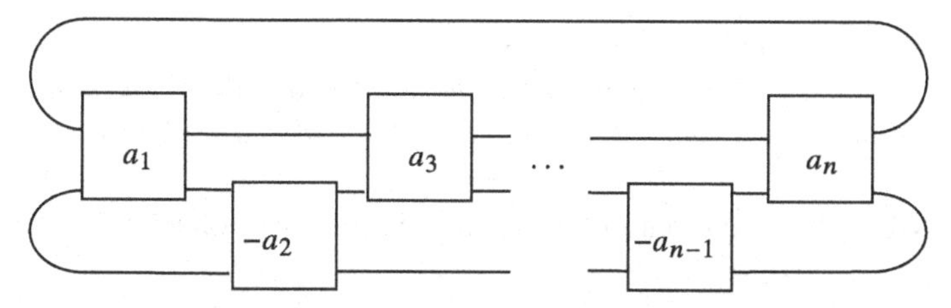

Figure 4.7 The two-bridge link $C(a_1, \dots, a_n)$. Here, a box with label k denotes k right-handed half-twists if $k > 0$ and k left-handed half-twists if $k < 0$.

where $\text{inj}_x(S^3 \setminus K)$ is the radius of a maximal embedded open ball about p. Then we have the following relationship between the knot signature and hyperbolic invariants, which was found with the help of machine learning; see Davies et al. [28]:

Theorem 4.53 *There is a constant c such that, for any hyperbolic knot K,*

$$|2\sigma(K) - slope(K)| \leq c vol(K) inj(K)^{-3}.$$

We know that $c \geq 0.23392$ and conjecture that $c \leq 0.3$.

For an in-depth introduction to hyperbolic knot theory, see Purcell [148].

4.5.5 Two-Bridge Links

Definition 4.54 We say that the link L in S^3 is *n-bridge* if it admits a diagram $\mathcal{D}$ such that the x-coordinate function has n local minima (and hence n local maxima), but it does not admit a diagram with $n-1$ local minima.

Schubert classified two-bridge links. They have also been called *rational links* by Conway, since they can be classified by rational numbers. The two-bridge link $C(a_1, \dots, a_n)$ is obtained by linearly plumbing $a_1, -a_2, a_3, \dots, (-1)^n a_n$-twisted unknotted bands; see Figure 4.7. We assign to $C(a_1, \dots, a_n)$ the rational number

$$\frac{p}{q} = \cfrac{1}{a_1 + \cfrac{1}{a_2 + \dots \cfrac{1}{a_{n-1} + \cfrac{1}{a_n}}}}.$$

Theorem 4.55 *Every two-bridge link is isotopic to some $C(a_1,\dots,a_n)$. Furthermore, two two-bridge links are isotopic if and only if the associated rational numbers p/q and p'/q' satisfy $p = p'$ and $q' \equiv q^{\pm 1} \mod p$.*

We will outline the proof of this result in Section 4.11 on branched double covers.

4.6 The Knot Group

Definition 4.56 Given a knot K in S^3, its *complement* is $S^3 \setminus K$, and its *exterior* is $S^3 \setminus N(K)$.

The complement and the exterior uniquely determine each other. Knot complements are considered more often when studying hyperbolic structures, while the exterior has the advantage of being compact. We now state a deep result of Gordon and Luecke [51]:

Theorem 4.57 *The isotopy class of a knot in S^3 is determined by the orientation-preserving homeomorphism type of its exterior.*

While the longitude of the knot is homologically determined by the half lives, half dies lemma (Lemma 3.12), there are infinitely many possible choices for the meridian. What Gordon and Luecke actually showed is that there is exactly one gluing of a solid torus $S^1 \times D^2$ to the knot exterior that gives S^3. The fact that the knots are in S^3 is important, since there are different knots in lens spaces with homeomorphic complements; see Bleiler–Hodgson–Weeks [10].

The fundamental group of the knot complement is a powerful knot invariant. A presentation, called the *Wirtinger presentation*, can be computed as follows:

Theorem 4.58 *Let $\mathcal{D}$ be a diagram of a knot K in S^3. Label the components of $\mathcal{D}$ by $a_1,\dots,a_c$ as we follow the orientation of K, where c is the number of crossings of $\mathcal{D}$, and we call crossing i the endpoint of the arc a_i. If a_j is the overstrand at crossing i, let*

$$w_i = \begin{cases} a_i a_j^{-1} a_{i+1}^{-1} a_j & \text{if crossing } i \text{ is positive,} \\ a_i a_j a_{i+1}^{-1} a_j^{-1} & \text{otherwise.} \end{cases}$$

Then

$$\langle\, a_1,\dots,a_c \mid w_1,\dots,w_c \,\rangle$$

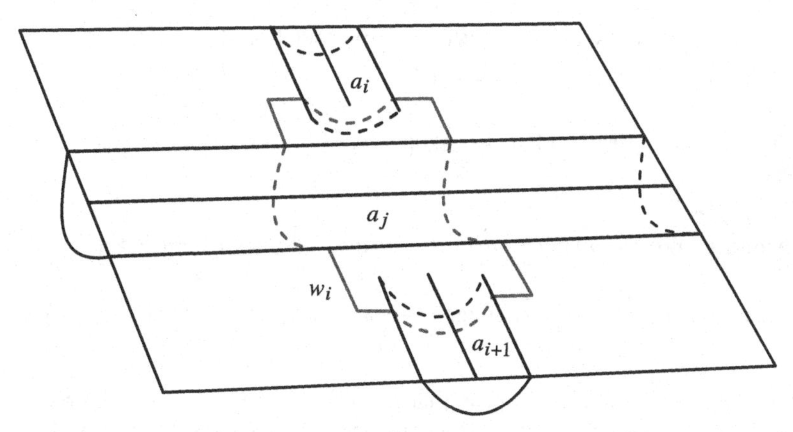

Figure 4.8 Illustration of the proof of the Wirtinger presentation. The zero-handle is the upper half-space, and the tunnels are one-handles attached below the plane. The two-cell over crossing i is attached from underneath along the curve w_i shown in grey.

is a presentation of $\pi_1(S^3 \setminus K)$. *In fact,* $S^3 \setminus K$ *is homotopy equivalent to a CW complex with a single zero-cell, one-cells* $a_1, \dots, a_c$*, two-cells attached along* $w_1, \dots, w_c$*, and a single three-cell.*

Proof It clearly suffices to construct the claimed CW decomposition. Let D^3 be the upper hemisphere of S^3, which we view as a zero-handle, with $\mathcal{D}$ lying on its boundary.

Let $N(a_i) \approx I \times [-1, 1]$ be a regular neighbourhood of a_i in S^2, where $a_i \approx I \times \{0\}$ and $a_i \times \{-1\}$ lies to the right of a_i. We attach a two-dimensional one-handle to D^3 along $I \times \{-1, 1\}$, which we view as a tunnel over a_i; see Figure 4.8. We orient its core from $I \times \{-1\}$ to $I \times \{1\}$. We repeat this for each arc a_i, after which we have a space homotopy equivalent to a wedge of c circles.

Over crossing i, we then attach a two-cell of the form $[-1, 1] \times [-1, 1]$. Its boundary goes over a_i in the positive direction, passes over the tunnel over a_j in the positive or negative direction depending on the sign of crossing i, goes over a_{i+1} in the negative direction, and finally goes over a_j in the opposite direction. Hence, the attaching map of the two-cell is precisely w_i.

Finally, we attach the three-dimensional three-handle corresponding to the lower hemisphere of S^3, and obtain a CW decomposition of $S^3 \setminus K$ of the required form. We remark that we can thicken the one-cells and two-cells to obtain a handle decomposition of the knot exterior. □

Exercise 4.59 Show using the Wirtinger presentation that $H_1(S^3 \setminus K) \cong \mathbb{Z}$. Use Alexander duality to compute $H_1(S^3 \setminus L)$ for a link L in S^3.

The following result of Papakyriakopoulos [144] is known as the *asphericity of knot complements*:

Theorem 4.60 *For a knot K in S^3, we have $\pi_n(S^3 \setminus K) = 0$ for $n > 1$.*

In other words, $S^3 \setminus K$ is an Eilenberg–MacLane space $K(\pi, 1)$ for $\pi = \pi_1(S^3 \setminus K)$.

Proof By contradiction, suppose that $\pi_2(S^3 \setminus K) \neq 0$. Then, by the Sphere Theorem (Theorem 3.10), there is an embedded two-sphere S in $S^3 \setminus K$ that is *not* null-homotopic. By the Schönflies theorem (Theorem 3.1), we have $S^3 \setminus S \approx B_1 \sqcup B_2$ for three-balls B_1 and B_2. As K is connected, it lies in one of B_1 and B_2, and the other ball shows that S is null-homotopic in $S^3 \setminus K$, which is a contradiction.

Let $\widetilde{E}$ denote the universal cover of the exterior E of K. By Exercise 4.59, the group $\pi_1(E)$ is infinite, and hence $\widetilde{E}$ is a noncompact three-manifold. It follows that $H_n(\widetilde{E}) = 0$ for $n > 2$. So, by the Hurewicz theorem, $\pi_n(E) \cong \pi_n(\widetilde{E}) = 0$ for $n > 2$. □

The knot group is not a complete knot invariant. For example, the square knot $3_1 \# -3_1$ and the granny knot $3_1 \# 3_1$ have isomorphic groups, where 3_1 is the right-handed and -3_1 is the left-handed trefoil. However, the knot group together with the subgroup $\pi_1(\partial M)$, called a *peripheral system*, is a complete knot invariant:

Theorem 4.61 *Let K and K' be knots in S^3 with exteriors M and M', respectively. If there is an isomorphism $\pi_1(M) \to \pi_1(M')$ that induces an isomorphism $\pi_1(\partial M) \to \pi_1(\partial M')$, then K and K' are equivalent. If K and K' are prime, then they are equivalent if and only if they have isomorphic groups.*

Sketch of proof A knot complement is an Eilenberg–MacLane space by Theorem 4.60. So, there is a homotopy equivalence $h\colon M \to M'$. The condition that $h_*\colon \pi_1(\partial M) \to \pi_1(\partial M')$ is an isomorphism ensures that h can be chosen to be a homotopy equivalence between ∂M and $\partial M'$, as T^2 is also an Eilenberg–MacLane space.

As M and M' are three-manifolds with boundary other than D^3, they are Haken, and we can apply Waldhausen's theorem, Theorem 3.14, to show that f

is homotopic to a homeomorphism using induction along the Haken hierarchy. To conclude that K and K' are equivalent knots, we invoke Theorem 4.57. □

We can compute the Alexander polynomial of a knot from a presentation of its fundamental group via *Fox calculus*: Given a free group F generated by $g_1, \ldots, g_n$, the *Fox derivatives* $\partial/\partial g_i \colon \mathbb{Z}[F] \to \mathbb{Z}[F]$ is a group homomorphism characterised by the following axioms:

(i) $\partial g_i/\partial g_j = \delta_{ij}$,
(ii) $\partial e/\partial g_i = 0$, and
(iii) $\partial(uv)/\partial g_i = \partial u/\partial g_i + u(\partial v/\partial g_i)$ for any $u, v \in F$.

Exercise 4.62 Deduce from the axioms that

$$\partial u^{-1}/\partial g_i = -u^{-1}(\partial u/\partial g_i)$$

for any $u \in F$.

Now suppose that $\langle g_1, \ldots, g_n \,|\, r_1, \ldots, r_m \rangle$ is a presentation of the knot group $G := \pi_1(S^3 \setminus K)$. Let F be the free group on $g_1, \ldots, g_n$, and

$$H := G/[G, G] \cong H_1(S^3 \setminus K) \cong \mathbb{Z}.$$

Consider the homomorphism $\phi\colon \mathbb{Z}[F] \to \mathbb{Z}[H]$, and write t for the generator of H. We form the matrix J whose (i, j)-th entry is $\phi(\partial r_i/\partial g_j)$ for $i \in \{1, \ldots, m\}$ and $j \in \{1, \ldots, n\}$. Then the greatest common divisor of all $(n-1) \times (n-1)$ minors of J agrees with the Alexander polynomial of K, up to multiplication by $\pm t^k$ for $k \in \mathbb{Z}$. (Recall that the Alexander polynomial is only well defined up to multiplication by t^k for $k \in \mathbb{Z}$.)

Exercise 4.63 Using Fox calculus, compute the Alexander polynomial of the trefoil knot.

4.7 Fibred Knots

Definition 4.64 A knot K in a three-manifold Y is said to be *fibred* if its complement $Y \setminus K$ is a fibre bundle over S^1 such that the closure of each fibre is a Seifert surface of K.

Example 4.65 The simplest example of a fibred knot is the unknot U, since $S^3 \setminus U \approx S^1 \times B^2$. The first non-trivial example is given by the trefoil knot. The figure eight knot is also fibred; we will shortly introduce a method that will allow us to easily prove this.

Example 4.66 The torus knot $T_{p,q}$ is fibred for any pair of relatively prime integers p and q. If we use the description of $T_{p,q}$ as the link of the singularity $w^p + z^q$ at the origin in $\mathbb{C}^2$ given in Exercise 4.42, then

$$\frac{w^p + z^q}{|w^p + z^q|} : S^3 \setminus K \to S^1$$

is a fibration. Indeed, the restriction of this map to the knot exterior is a proper submersion, which is hence a fibre bundle by Ehresmann's fibration lemma (Exercise 1.41).

More generally, suppose that $f: \mathbb{C}^2 \to \mathbb{C}$ is a complex polynomial in variables w and z such that $f(\underline{0}) = 0$, $\frac{\partial f}{\partial w}(\underline{0}) = 0$, $\frac{\partial f}{\partial z}(\underline{0}) = 0$, and $\underline{0} \in \mathbb{C}^2$ is an isolated common zero of these functions. Then $V := f^{-1}(0)$ is an algebraic variety with an isolated singularity at $\underline{0}$. The *link of the singularity* is the link $L := V \cap S^3_\varepsilon$ for some $\varepsilon > 0$ small. Then L is a fibred link with fibration $g/|g|: S^3 \setminus L \to S^1$, where $g := f|_{S^3 \setminus L}$. This is called the *Milnor fibration* of the singularity. For more detail, see Milnor [112].

The following result of Stallings [15] gives a complete characterisation of fibred knots:

Theorem 4.67 *Let K be a knot in S^3. Then K is fibred if and only if the commutator subgroup of $\pi_1(S^3 \setminus K)$ is finitely generated.*

Proof Let E be the knot exterior. First, suppose that K is fibred with fibre S. From the homotopy long exact sequence of the fibration $S \to E \to S^1$, the sequence

$$0 \to \pi_1(S) \to \pi_1(E) \xrightarrow{h} \mathbb{Z}$$

is exact. As $\mathbb{Z}$ is Abelian, the map h factors through

$$\pi_1(E)/[\pi_1(E), \pi_1(E)] \cong H_1(E) \cong \mathbb{Z}.$$

Since every homomorphism $\mathbb{Z} \to \mathbb{Z}$ is injective,

$$\ker(h) = [\pi_1(E), \pi_1(E)] \cong \pi_1(S),$$

which is finitely generated.

Conversely, suppose that $[\pi_1(E), \pi_1(E)]$ is finitely generated. Let S be a minimal genus Seifert surface for K, and let M be the three-manifold obtained by cutting the knot exterior E along S. We denote by S_+ and S_- the subsurfaces of ∂M corresponding to the two sides of S. Since $[\pi_1(E), \pi_1(E)]$ is finitely generated, the inclusion maps

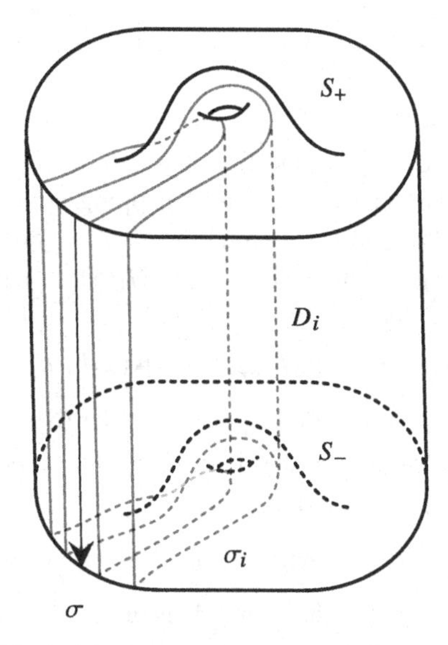

Figure 4.9 An illustration of the proof of Stallings' theorem.

$$\pi_1(S_\pm) \xrightarrow{\sim} \pi_1(M) \tag{4.3}$$

are isomorphisms; see Burde–Zieschang [15].

Let σ be an arc on ∂M obtained by cutting a meridian of K along S, with $\sigma(0) \in S_+$ and $\sigma(1) \in S_-$; see Figure 4.9. By Equation (4.3), for every loop $\gamma \in \pi_1(S_-, \sigma(1))$, there is a loop $f_*(\gamma) \in \pi_1(S_+, \sigma(0))$ that is homotopic to $\sigma\gamma\sigma^{-1}$. Then f_* is induced by a homeomorphism $f: S_- \to S_+$ by Nielsen's theorem [131], which states that every automorphism of the fundamental group of a compact surface can be realised by a homeomorphism. (This might be viewed as a two-dimensional analogue of Waldhausen's theorem.) We can further assume that $f(\sigma(1)) = \sigma(0)$.

Choose a set of curves $\gamma_1, \dots, \gamma_{2g(S_-)}$ on S_- based at $\sigma(1)$ such that cutting S_- along them results in a disk. Then we can slightly isotope the curves $\sigma\gamma_i\sigma^{-1} f(\gamma_i)^{-1}$ on ∂M to disjoint simple closed curves σ_i. As σ_i is null-homotopic in M, it bounds a disk D_i by Dehn's lemma (Theorem 3.6). If we cut M along the disks D_i, the resulting manifold M' is a three-disk by the Schönflies theorem (Theorem 3.1) as $\partial M' \approx S^2$. Let $S'_\pm$ be the result of cutting $S_\pm$ along the D_i. Then $S'_\pm \approx D^2$, so we can identify (M', S'_-, S'_+) with the product $(D^2 \times I, D^2 \times \{0\}, D^2 \times \{1\})$, endowing it with a fibration over I that glues together to a fibration of M over I with fibres including S_+ and S_-. Finally, gluing S_+ to S_- gives the desired fibration of E over S^1. □

Proposition 4.68 *Let K be a fibred knot in S^3. Then the Alexander polynomial $\Delta_K(t)$ is monic.*

Proof Let S be a fibre surface, and fix a basis $\mathcal{B} = \{b_1, \dots, b_{2g}\}$ of $H_1(S)$. Then we can write the exterior of K as a mapping torus S_φ, where φ is an automorphism of $(S, \partial S)$ and is called the *monodromy* of the fibration. Let V be the Seifert matrix and M the matrix of $\varphi_* \colon H_1(S) \to H_1(S)$ with respect to $\mathcal{B}$. Note that

$$\mathrm{lk}(b_i, b_j^+) = \mathrm{lk}(b_i^+, \varphi_* b_j) = \mathrm{lk}(\varphi_* b_j, b_i^+),$$

by translating b_i and b_j^+ along the mapping cylinder $(S \times I)/_{(x,1)\sim(\varphi(x),0)}$ until b_j^+ reaches $S \times \{1\}$. This implies that $V = M^T V^T$ and V is invertible, so $M = V^{-1} V^T$. Hence, the Alexander polynomial

$$\Delta_K(t) = \det(V - tV^T) = \det(V)\det(I - tV^{-1}V^T) = \det(I - tM).$$

As φ is an automorphism of S, the matrix M is invertible over $\mathbb{Z}$, so the leading coefficient of $\Delta_K(t)$ is $\det(M) = \pm 1$. □

If K is fibred, the surface obtained by taking the closure of a fibre of the fibration $S^3 \setminus K \to S^1$ is a minimal genus Seifert surface of K, and is unique.

Gabai introduced sutured manifolds to give a simple method to check whether a knot is fibred. This formalises the method applied in the proof of Theorem 4.67.

Definition 4.69 A *sutured manifold* is a pair (M, γ), where M is a compact, oriented three-manifold with boundary, and γ is a collection of thickened, oriented simple closed curves in ∂M that divide ∂M into two subsurfaces $R_+(\gamma)$ and $R_-(\gamma)$ that meet along γ. Furthermore, $R_+(\gamma)$ is oriented as ∂M, while $R_-(\gamma)$ is oriented as $-\partial M$, and γ is oriented as the boundary of both $R_+(\gamma)$ and $R_-(\gamma)$.

Given a knot K in an oriented three-manifold Y and a Seifert surface S for K, the *sutured manifold complementary to S* is given by the pair (M, γ), where $M = Y \setminus N(S)$, and γ is a thickening of K in ∂M. The positive side of S is $R_+(\gamma)$ and the negative side is $R_-(\gamma)$.

We say that (M, γ) is a *product sutured manifold* if $M = R \times I$ and $\gamma = \partial R \times I$ for some compact, oriented surface R. Here $R_-(\gamma) = R \times \{0\}$ and $R_+(\gamma) = R \times \{1\}$.

Proposition 4.70 *The knot K in S^3 is fibred with fibre S if and only if the sutured manifold complementary to S is a product.*

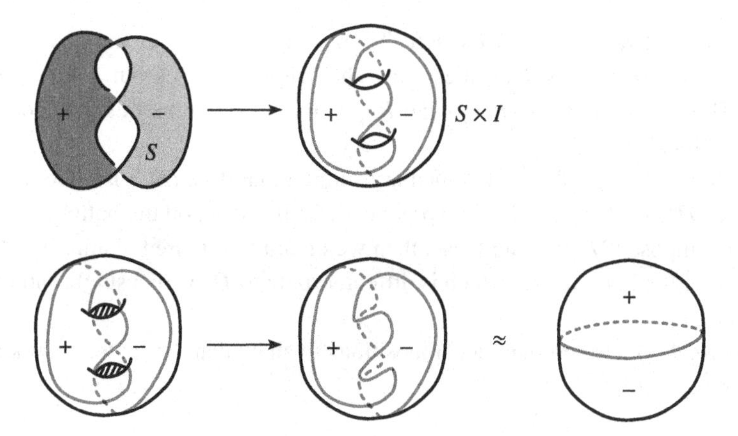

Figure 4.10 A proof that the trefoil knot is fibred using sutured manifolds.

Proof This is a straightforward corollary of the definitions. □

Given a sutured manifold (M,γ), a *product disk* in (M,γ) is a properly embedded disk $D \subset M$ such that $|D \cap \gamma| = 2$. We can decompose (M,γ) along D by taking $M' := M \setminus N(D)$, and γ' is obtained by reconnecting the ends of $\gamma \setminus N(D)$ along $D \times \{-1\}$ and $D \times \{1\}$, where we identify $N(D)$ with $D \times [-1,1]$. Then

$$(M,\gamma) \overset{D}{\rightsquigarrow} (M',\gamma')$$

is called a *product decomposition*. The following result is straightforward:

Proposition 4.71 *The connected sutured manifold (M,γ) is a product if and only if it admits a sequence of product decompositions terminating at a three-ball with a single suture on it.*

This gives us a practical method for showing that a knot K in S^3 is fibred: Choose a Seifert surface S for K, usually using Seifert's algorithm. Then consider the product sutured manifold $(S \times I, \partial S \times I)$ by thickening S and considering K as the suture. We then look for product disks in the *complement* of $(S \times I, \partial S \times I)$. Decomposing the complement along such a disk D amounts to adding $D \times [-1,1]$ to $(S \times I, \partial S \times I)$ and reconnecting the sutures along the boundary. If, at the end of this process, we end up with D^3 with a single suture, then K is fibred, and S is a minimal genus Seifert surface that is a fibre.

Example 4.72 Figure 4.10 illustrates the above procedure in the case of the right-handed trefoil knot. Consider the Seifert surface S shown in the upper left. Its positive side is on the left. Thickening S results in the genus two handlebody $H := S \times I$ on the right. The original knot is our suture γ on ∂H. This divides ∂H into $R_+(\gamma)$ and $R_-(\gamma)$. By construction, this is a product sutured manifold.

We now focus on the complementary sutured manifold (M, γ), where $M := S^3 \setminus \text{Int}(H)$. We have shaded two product disks in (M, γ) on the bottom left. If we decompose (M, γ) along these, then we obtain the sutured manifold in the middle of the bottom row, which is diffeomorphic to D^3 with a single suture.

Exercise 4.73 Use product decompositions to show that the figure eight knot is fibred.

4.8 The Jones and HOMFLY Polynomials

Vaughan Jones [71] introduced a novel invariant of links in 1984 using von Neumann algebras, which was the first new knot polynomial after the Alexander polynomial. It admits a much simpler definition due to Kauffman [78]. Unlike the Alexander polynomial, its relationship with the geometry of the link is much less clear. It is conjectured to detect the unknot. It was the main tool for proving Tait's conjectures (Theorem 4.37). It admits a categorification called Khovanov homology – a homology theory whose graded Euler characteristic is the Jones polynomial – that has recently been shown to detect the unknot by Kronheimer and Mrowka [91].

Let L be a link in S^3. The Jones polynomial $V_L(t)$ is a Laurent polynomial in $\mathbb{Z}[t^{1/2}, t^{-1/2}]$ characterised by $V_U(t) = 1$ and the oriented skein relation

$$(t^{1/2} - t^{-1/2})V_{L_0}(t) = t^{-1}V_{L_+}(t) - tV_{L_-}(t),$$

where L_+, L_-, and L_0 are link diagrams that agree outside a single positive crossing c of L_+. The link L_- is obtained from L_+ by changing c to a negative crossing, and L_0 is the oriented resolution of c; see the top row of Figure 4.11. When the number of components of L is odd, then $V_L(t) \in \mathbb{Z}[t, t^{-1}]$.

It is not clear from this description that there exists such a polynomial and that it is unique. We now give the definition of the Jones polynomial using the Kauffman bracket, which satisfies a different type of skein relation.

Definition 4.74 Let L be an unoriented link diagram in S^2. Then its *Kauffman bracket* $\langle L \rangle \in \mathbb{Z}[A, A^{-1}]$ is characterised by the following relations:

(i) $\langle \bigcirc \rangle = 1$,
(ii) $\langle L \sqcup \bigcirc \rangle = -(A^2 + A^{-2})\langle L \rangle$, and

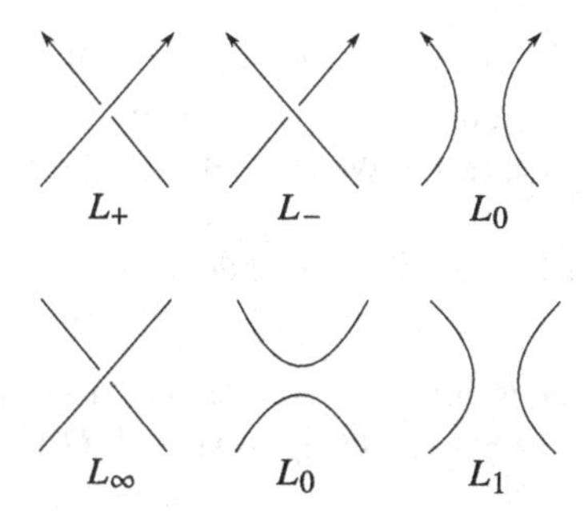

Figure 4.11 The top row shows the three links featuring in the oriented skein relation, and the bottom row the links in the unoriented skein relation.

(iii) if the diagrams L_∞, L_0, and L_1 are as in the bottom row of Figure 4.11, then $\langle L_\infty \rangle = A^{-1}\langle L_0 \rangle + A\langle L_1 \rangle$ (*unoriented skein relation*).

If a diagram L has n crossings, there are two ways of resolving each, giving rise to 2^n unlink diagrams. The Kauffman bracket $\langle L \rangle$ is then recursively obtained as a combination of the bracket polynomials of these unlinks. A c-component unlink has bracket $(-A^2 - A^{-2})^{c-1}$. To see that this is independent of the order of resolutions, one checks that the result is unchanged if we swap the order of two adjacent crossings. The Kauffman bracket is not quite a link invariant:

Lemma 4.75 *Let L be a link diagram in S^2.*

(i) *Suppose that L' is obtained from L via an R1 move. Then either the zero-resolution or the one-resolution of L' at the new crossing splits off an unknot component. In the former case, $\langle L' \rangle = -A^{-3}\langle L \rangle$, and, in the latter, $\langle L' \rangle = -A^3\langle L \rangle$.*

(ii) *If L' is obtained from L by an R2 or R3 move, then $\langle L \rangle = \langle L' \rangle$.*

Proof Consider part (i), and assume that the one-resolution splits off an unknot. If we set $L_\infty := L'$ and apply the unoriented skein relation to it, then $L_0 = L$ and $L_1 = L \sqcup \bigcirc$, so we obtain that

$$\langle L' \rangle = A^{-1}\langle L \rangle + A\langle L \sqcup \bigcirc \rangle = A^{-1}\langle L \rangle - A(A^2 + A^{-2})\langle L \rangle = -A^3\langle L \rangle,$$

as claimed. The case when the zero-resolution splits off an unknot is analogous.

Invariance under Reidemeister move R2 is obtained by applying the skein relation to one of the two new crossings, followed by part (i).

Finally, R3-invariance can be seen by applying the skein relation to the crossing that does not lie on the moving strand before and after the R3 move, and applying R2-invariance twice. We leave the details as an exercise. □

It is immediate that the following is well defined:

Definition 4.76 Let L be a link in S^3 and choose a diagram in S^2. Then the *Jones polynomial* $V_L(t)$ is obtained from $(-A)^{-3w(L)}\langle L\rangle$ by substituting $t^{1/2} = A^{-2}$.

Unlike the Alexander polynomial, the Jones polynomial is hard to compute, and current algorithms are exponential-time in the crossing number n of the diagram (corresponding to the 2^n resolutions).

It is a simple consequence of the preceding characterisation of the Jones polynomial that $V_{-K}(t) = V_K(t^{-1})$. Hence, if $V_K(t) \neq V_K(t^{-1})$, then the knot is *chiral*; that is, $K \neq -K$. For example, $V_{3_1}(t) = t+t^3-t^4$ is not a symmetric polynomial, hence the left-handed and right-handed trefoils are inequivalent knots. As for the Alexander polynomial, the Jones polynomial is also alternating for alternating links, as shown by Thistlethwaite.

It is clear from the definition that the span of the Jones polynomial gives a lower bound on the crossing number of the link diagram (and hence on the crossing number of the link). For an alternating, reduced link diagram, Lickorish and Millett [99] showed that the span is equal to the crossing number of the diagram. This implies the first Tait conjecture (Theorem 4.37).

Remark 4.77 The *N-coloured Jones polynomial* $J_{K,N}(t)$ for $N \in \mathbb{Z}_{>0}$ is the Jones polynomial of the untwisted N-cable of K. In particular, $J_{K,1}(t) = V_K(t)$. The *Kashaev invariant* of K is defined as

$$\langle K\rangle_N := \lim_{q \to e^{2\pi i/N}} \frac{J_{K,N}(q)}{J_{U,N}(q)}.$$

The *volume conjecture* is the main motivating problem in the area. This states that, for a hyperbolic knot K, the quantity

$$\lim_{N\to\infty} \frac{2\pi \log|\langle K\rangle_N|}{N}$$

agrees with the hyperbolic volume of K. This is based on computational evidence due to Kashaev, who also verified the conjecture for the knots 4_1, 5_2, and 6_1.

The *HOMFLY polynomial* is a two-variable integral Laurent polynomial in variables ℓ and m that extends both the Alexander and Jones polynomials. It is named after Hoste, Ocneanu, Millett, Freyd, Lickorish, and Yetter [39]. It was independently also discovered by Przyticki and Traczyk [147], and is hence also sometimes called the HOMFLY-PT polynomial. It can be defined using skein relations by requiring that $P_U = 1$ and

$$\ell P_{L_+} + \ell^{-1} P_{L_-} + m P_{L_0} = 0.$$

If L is the split union of L_1 and L_2, then we define

$$P_L = -\frac{\ell + \ell^{-1}}{m} P_{L_1} P_{L_2}.$$

Since $P_K(\ell, m) = P_{-K}(\ell^{-1}, m)$, the HOMFLY polynomial can also be used to show a knot is chiral.

One can also obtain the HOMFLY polynomial using the skein relation

$$a P_{L_+} - a^{-1} P_{L_-} = z P_{L_0}.$$

If we substitute $a = 1$ and $z = t^{1/2} - t^{-1/2}$, we recover the Alexander polynomial. The substitution $a = t^{-1}$ and $z = t^{1/2} - t^{-1/2}$ gives the Jones polynomial.

4.9 The Concordance Group

The notion of knot concordance was introduced by Fox and Milnor [36]. For a more detailed survey, see the work of Livingston [103].

Definition 4.78 Knots K_0 and K_1 in S^3 are smoothly (respectively topologically) *concordant* if there is a smooth (respectively locally flat) embedding $C\colon I \times S^1 \hookrightarrow I \times S^3$ such that $C|_{\{i\} \times S^1} = K_i$ for $i \in \{0, 1\}$.

Concordance is analogous to the notion of pseudoisotopy for diffeomorphisms, extended to embeddings. If the embedding C is level-preserving, the knots K_0 and K_1 are isotopic.

Concordance is an equivalence relation on the set of knots in S^3. A knot is smoothly concordant to the unknot if and only if it is slice; that is, $g_4(K) = 0$. Smooth (respectively topological) concordance classes of knots form an Abelian group called the smooth (respectively topological) *concordance group*, denoted $\mathcal{C}$ (respectively $\mathcal{C}_{\text{top}}$), where the group operation is connected sum, the identity element is the class of the unknot, and the inverse of K is $-K$. Indeed, if $p \in K$ and $a = I \times \{p\}$, then $(I \times S^3) \setminus N(a) \approx D^4$ after smoothing the corners, and $(I \times K) \setminus N(a)$ is a two-disk with boundary $K \# -K$. Hence K_0 and K_1 are concordant if and only if $K_0 \# -K_1$ is slice.

We have already encountered two concordance invariants, namely the signature and the Arf invariant. In fact, both of them are homomorphisms from the concordance group. Indeed, they are additive under connected sums and vanish on slice knots; see Propositions 4.28 and 4.32. Further concordance homomorphisms can be constructed using the Seifert form, Heegaard Floer homology, and Khovanov homology. These allow one to show that the concordance group is infinitely generated.

If a knot is achiral (i.e., $K = -K$), then it is a two-torsion element of C. Fox [34] proved that the figure eight knot 4_1 is achiral but not slice using the Alexander polynomial. Murasugi [126] used the knot signature to show that the trefoil knot has infinite order in C. Levine [95][96] constructed an epimorphism

$$\phi\colon C \to \mathbb{Z}^\infty \oplus \mathbb{Z}_2^\infty \oplus \mathbb{Z}_4^\infty,$$

which we will discuss shortly. Casson and Gordon [17][18] proved, using a new invariant, that $\ker(\phi) \neq \{0\}$. Jiang [69] found a $\mathbb{Z}^\infty$ subgroup of $\ker(\phi)$. Livingston [102] showed that $\mathbb{Z}_2^\infty < \ker(\phi)$. It is an open question whether there is torsion of order other than two or four in C.

The next result of Freedman [37] follows from his classification of topological four-manifolds. It is the main method for showing a knot is topologically slice.

Theorem 4.79 *Let K be a knot in S^3. If $\Delta_K(t) = 1$, then K is topologically slice.*

That the topological and smooth concordance groups differ substantially follows from the work of Donaldson, Seiberg–Witten, and Ozsváth–Szabó. For example, if K is the untwisted Whitehead double of the trefoil, then it is not smoothly slice (which can be shown, for example, using the Rasmussen s-invariant in Khovanov homology or the Ozsváth–Szabó τ-invariant in knot Floer homology; see Section 5.3); but $\Delta_K(t) = 1$ by Exercise 4.49, and so it is topologically slice.

Conjecture 4.80 A knot is smoothly slice if and only if its Whitehead double is smoothly slice.

Cochran, Orr, and Teichner [23] constructed a descending filtration on C indexed by integers and half-integers:

$$\cdots \subsetneq \mathcal{F}_2 \subsetneq \mathcal{F}_{1.5} \subsetneq \mathcal{F}_1 \subsetneq \mathcal{F}_{0.5} \subsetneq \mathcal{F}_0 \subsetneq C.$$

Here, each inclusion is proper. Furthermore, $\mathcal{F}_0$ is the collection of knots with vanishing Arf invariant, $\mathcal{F}_{0.5} = \ker(\phi)$, and every knot in $\mathcal{F}_{1.5}$ has vanishing Casson–Gordon invariant.

We will now give an overview of the work of Levine, and explain the construction of the homomorphism ϕ. Let S be a Seifert surface for the knot K in S^3. Recall that one can associate to it a Seifert form

$$V_S\colon H_1(S) \times H_1(S) \to \mathbb{Z}.$$

Furthermore, if V is a Seifert matrix corresponding to a free generating set of $H_1(S)$, then $V - V^T$ represents the intersection form on $H_1(S)$ and is hence unimodular. This motivates the following definition:

Definition 4.81 We say that (A, V) is an *abstract Seifert pairing* if A is a finitely generated, free Abelian group, and

$$V\colon A \times A \to \mathbb{Z}$$

is a bilinear pairing such that $V - V^T$ is unimodular.

The abstract Seifert pairing (A, V) is *metabolic* if $A = A_1 \oplus A_2$, such that $A_1 \cong A_2$, and $V(x, y) = 0$ for every x, $y \in A_1$. In this case, we call A_1 a *metaboliser*.

Proposition 4.82 *Let K be a knot in S^3 with Seifert surface S. If K is slice, then V_S is metabolic.*

Proof This follows from our proof of Proposition 4.28. More specifically, if D is a slice disk for K, then we choose a three-manifold $R \subset D^4$ bounding $S \cup D$ and set

$$A_1 := \ker(H_1(S) \to H_1(R)).$$

Then $\mathrm{rk}(A_1) = \mathrm{rk}(H_1(S))/2$ by the half lives, half dies lemma. If x, $y \in A_1$, then they bound two-chains c_x and c_y in R. By pushing c_y off R in D^4 in the normal direction, we see that the linking number of x and the push-off of y vanishes. □

Corollary 4.83 *Let K_0 and K_1 be concordant knots with Seifert forms V_0 and V_1, respectively. Then $V_0 \oplus -V_1$ is metabolic.*

Definition 4.84 The abstract Seifert forms V_0 and V_1 are called *algebraically concordant* if $V_0 \oplus -V_1$ is metabolic.

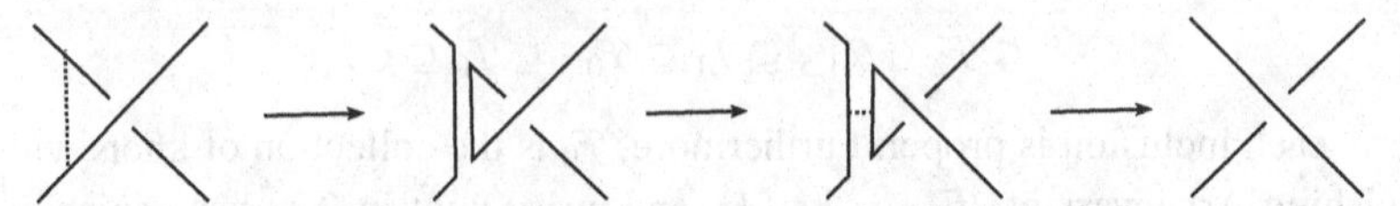

Figure 4.12 Realising a crossing change via two band moves. The first and third arrows are band moves, where the bands are attached along the dotted lines with blackboard framing. The middle arrow is an isotopy.

Proposition 4.85 *The algebraic concordance classes of abstract Seifert forms form a group with respect to direct sum, where the identity element is the rank zero $\mathbb{Z}$-module.*

We denote the algebraic concordance group by $\mathcal{G}$.

Theorem 4.86 *The map $\phi\colon \mathcal{C} \to \mathcal{G}$ that assigns to a knot K the algebraic concordance class of a Seifert form is a surjective homomorphism.*

Proof That the map ϕ is well defined follows from the preceding discussion. To show that it is surjective, one explicitly constructs a Seifert surface S for each abstract Seifert form V such that $V_S = V$. This can be realised by attaching twisted bands to D^2 that link according to V. □

Levine proved that $\mathcal{G} \cong \mathbb{Z}^\infty \oplus \mathbb{Z}_2^\infty \oplus \mathbb{Z}_4^\infty$. The group $\ker(\phi)$ consists of so-called *algebraically slice* knots.

Concordance is a special case of knot cobordism:

Definition 4.87 The knots K_0 and K_1 in S^3 are *cobordant* if there is a properly embedded, compact, oriented surface S in $I \times S^3$ such that $\partial S = -(\{0\} \times K_0) \cup (\{1\} \times K_1)$.

So two knots are concordant if there is a genus zero cobordism between them. We can visualise a knot cobordism using Morse theory as a sequence of Reidemeister moves, adding or removing unknot components, and oriented band moves. Such a sequence is sometimes called a *movie*.

Any two knots in S^3 are cobordant, since we can realise a crossing change using two band moves (see Figure 4.12), and we can convert any knot to the unknot using crossing changes. The genus of a connected knot cobordism S can be computed using the formula

$$g(S) = -\frac{\chi(S)}{2} = \frac{b-u}{2},$$

where u is the number of unknot components that appear or disappear, and b is the number of bands.

Definition 4.88 The knot K is *ribbon* if it is slice, and there is a slice disk such that the radial function on D^4 has no local maxima along it. Such a slice disk is called a *ribbon disk.*

In terms of movies, a ribbon disk consists of b band moves that result in a $(b + 1)$-component unlink. The following famous conjecture is due to Fox and is known as the slice-ribbon conjecture:

Conjecture 4.89 Every slice knot is ribbon.

Gordon extended the notion of ribbon disks to concordances:

Definition 4.90 We say that there is a *ribbon concordance* from K_0 to K_1, and write $K_0 \leq K_1$, if there is a concordance C between them such that the projection $S^3 \times I \to I$ does not admit local maxima.

It is clear that the relation $\leq$ is reflexive and transitive. By the following result of Agol [5], the relation $\leq$ is antisymmetric, and hence a partial order:

Theorem 4.91 *If $K_0 \leq K_1$ and $K_1 \leq K_0$, then $K_0 = K_1$.*

We have $U \leq K$ if and only if K is ribbon. We can always isotope a concordance C from K_0 to K_1 such that one first encounters index zero critical points, followed by fusion saddles, then fission saddles, and finally index two critical points. If K is the knot obtained after passing the index zero critical points and the fusion saddles, then $K \geq K_0$ and $K \geq K_1$.

The following result is due to Gordon [49]:

Proposition 4.92 *If C is a ribbon concordance from K_0 to K_1, then the map*

$$\pi_1(S^3 \setminus K_i) \to \pi_1((I \times S^3) \setminus C)$$

is injective for $i = 0$ and surjective for $i = 1$.

Proof Let us write $Y := (I \times S^3) \setminus N(C)$ and $E_i := (\{i\} \times S^3) \setminus N(K_i)$ for $i \in \{0, 1\}$. The claim for $i = 1$ is simple: There is a handle decomposition of Y relative to $I \times E_1$ with only two-handles (corresponding to saddles of C)

and three-handles (corresponding to local minima of C). A two-handle adds a relation to the fundamental group, while a three-handle leaves it unchanged.

We now show that the map $\varphi\colon \pi_1(E_0) \to \pi_1(Y)$ is injective. There is a handle decomposition of Y relative to $I \times E_0$ with only one-handles and two-handles. Since the embedding $E_0 \hookrightarrow Y$ induces an isomorphism on homology, the two-handles cancel the one-handles homologically. Hence

$$\pi_1(Y) \cong (\pi_1(E_0) * F)/\langle r_1, \dots, r_n \rangle,$$

where F is a free group on $x_1, \dots, x_n$ corresponding to the one-handles. Furthermore,

$$\det(\varepsilon_i(r_j)) = \pm 1,$$

where $\varepsilon_i(r_j)$ is the exponent sum of x_i in r_j.

By the work of Thurston, $\pi_1(E_0)$ is residually finite. Hence, if $\varphi(z) = 1$ but $z \neq 1$, then there is an epimorphism $e\colon \pi_1(E_0) \to G$ for a finite group G such that $e(z) \neq 1$. Then there is an epimorphism from $\pi_1(Y)$ to

$$H := (G * F)/\langle e(r_1), \dots, e(r_n) \rangle.$$

We claim that the induced map $h\colon G \to H$ is injective. Note that $\varepsilon_i(r_j) = \varepsilon_i(e(r_j))$. We now need a result of Gerstenhaber and Rothaus [45]:

Lemma 4.93 *Let G be a finite group, and $w_1, \dots, w_n$ words in $x_1, \dots, x_n$ and elements of G such that* $\det(\varepsilon_i(w_j)) \neq 0$. *Then there is a finite extension K of G containing elements $X_1, \dots, X_n$ such that $w_i(X_1, \dots, X_n) = 1$ for $i \in \{1, \dots, n\}$.*

If we apply the above lemma with $w_i = e(r_i)$, we obtain a finite extension $f\colon G \to K$ that factors through the map h; that is, the following diagram is commutative:

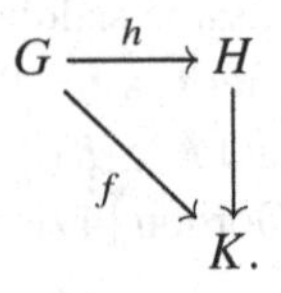

Since f is injective, so is h, as claimed. As $e(z) \neq 1$, we have $h(e(z)) \neq 1 \in H$. But $h \circ e$ factors through the map $\varphi\colon \pi_1(E_0) \to \pi_1(Y)$, which contradicts our assumption that $\varphi(z) = 1$. Hence φ is injective. □

Motivated by the preceding result, the concordance C from K_0 to K_1 is called *homotopy ribbon* if it satisfies the conclusions of Proposition 4.92.

If C is a ribbon concordance, then one obtains a handle decomposition of $(I \times S^3) \setminus N(C)$ using only one-handles corresponding to the local minima and two-handles corresponding to the saddles of the projection $I \times S^3 \to I$. We say that C is a *strong homotopy ribbon concordance* if $(I \times S^3) \setminus N(C)$ admits such a handle decomposition.

Every ribbon concordance is strongly homotopy ribbon, and every strongly homotopy ribbon concordance is homotopy ribbon. It is not known if these inclusions are strict.

We conclude the section with the following striking result of Zemke [180], answering a conjecture of Gordon, whose proof uses knot Floer homology:

Theorem 4.94 *If* $K_0 \leq K_1$, *then* $g(K_0) \leq g(K_1)$.

4.10 Dehn Surgery

We have already encountered surgery on an arbitrary n-manifold M along a framed, embedded k-sphere S; see Definition 1.58. This described the outgoing boundary of the handle attachment to $I \times M$ along $\{1\} \times S$. Concretely, one removes a regular neighbourhood $N(S) \approx S^k \times D^{n-k}$ of S, and glues in $D^{k+1} \times S^{n-k-1}$ along $\partial N(S) = S^k \times S^{n-k-1}$ using the framing.

When we have a framed knot K in a three-manifold Y, we are in the special situation that $\partial N(K) \approx S^1 \times S^1$ has a large mapping class group, namely $\mathrm{SL}(2, \mathbb{Z})$, and we are not restricted to gluing via a framing. As in the case of lens spaces, any automorphism of $S^1 \times S^1$ that preserves $S^1 \times \{x\}$ for some $x \in S^1$ extends to $D^2 \times S^1$. Hence, the result of gluing $D^2 \times S^1$ to $Y \setminus N(K)$ is determined, up to diffeomorphism, by the homology class of the image of $S^1 \times \{x\}$ in $\partial N(K)$.

Definition 4.95 Let K be a knot in the three-manifold Y. If we glue $D^2 \times S^1$ to $Y \setminus N(K)$ such that the image of $S^1 \times \{x\}$ is a slope $\alpha \in H_1(\partial N(K))$, then we write $Y_\alpha(K)$ for the result and call it the *Dehn surgery* along K with slope α.

If Y is a homology three-sphere, then K is null-homologous in Y, and we have a well-defined longitude $\ell \in H_1(\partial N(K))$, which is the class null-homologous in $Y \setminus N(K)$. This exists by the half lives, half dies lemma (Lemma 3.12). If we write $m \in H_1(\partial N(K))$ for the class of the meridian of K, the result of the Dehn surgery is determined by an extended rational number $p/q \in \mathbb{Q} \cup \{\infty\}$, where p and q are coprime, and the image of $S^1 \times \{x\}$ is $pm + q\ell \in H_1(\partial N(K))$. Gluing along a framing of K corresponds to $q = 1$; that is, to integer surgery. When $p/q \notin \mathbb{Z}$, the surgery does not correspond to a handle attachment.

More generally, one can consider Dehn surgery along a framed link L in a three-manifold Y whose components $L_1, \dots, L_n$ are labelled by extended rational numbers $p_1/q_1,\dots,p_n/q_n$. Indeed, the framing determines a canonical longitude ℓ_i of L_i for $i \in \{1,\dots,n\}$, and if we write m_i for the meridian of L_i, then we use the surgery slope $p_i m_i + q_i \ell_i$ for L_i. As explained earlier, when Y is a homology three-sphere, L has a canonical framing. We have the following result of Lickorish [97] and Wallace [176]:

Theorem 4.96 *Every closed, connected, and oriented three-manifold Y can be obtained by Dehn surgery along a link in S^3. Furthermore, the surgery coefficients can be chosen to be integers.*

Proof The oriented cobordism group Ω_3^{SO} of three-manifolds vanishes; see Equation (1.3) in Section 1.7. Hence Y bounds an oriented four-manifold X. Choose a handle decomposition of X. By cancelling zero-handles and one-handles, we can assume that X has a single zero-handle. Similarly, we can assume that X has no four-handles. Given a one-handle, we can cap off its core in the zero-handle to obtain an embedded S^1. If we perform surgery on X along it, we replace $S^1 \times D^3$ with $D^2 \times S^2$, which corresponds to attaching a five-dimensional two-handle to $I \times X$. (In the language of Kirby calculus, which we will discuss in Section 6.1, this corresponds to replacing a dotted circle representing a one-handle with a zero-framed two-handle.) If we turn the handle decomposition upside down, we can similarly replace three-handles with two-handles. The result of these surgeries admits a handle decomposition with a single zero-handle and some two-handles. This means that M can be obtained by performing integer surgery along a link in S^3. □

Definition 4.97 Given a three-manifold M with a torus boundary component T, we can glue $S^1 \times D^2$ to M along T. This operation is called *Dehn filling*.

So one can think of Dehn surgery as removing a neighbourhood of a knot, and Dehn filling the resulting manifold with torus boundary.

It is currently an active area of research to determine which surgeries on knots in S^3 give rise to lens spaces. The class of these knots are described by the *Berge conjecture*.

Definition 4.98 We say that the knot K in S^3 is a *Berge knot* (or *doubly primitive knot*) if there is a genus two Heegaard surface S in S^3 such that $K \subset S$, and each of the two handlebodies bounded by S has a meridian that intersects K in a single point.

The following conjecture is due to Berge and Gordon, and is problem 1.78 in Kirby's list of problems [86]:

Conjecture 4.99 A knot K in S^3 admits a lens space surgery if and only if it is a Berge knot.

Ni [129][130] showed that if K admits a lens space surgery, then it is fibred. Furthermore, Greene [52] proved that the set of lens spaces that arise via surgery on a knot in S^3 is precisely the set of lens space surgeries on Berge knots. Both of these results rely on Heegaard Floer homology, which we will overview in Section 5. However, it is still unknown whether there are non-Berge knots that admit lens space surgeries.

Definition 4.100 Let K be a knot in an oriented three-manifold Y, and α and β two slopes on $\partial N(K)$. Then we say that the surgeries along (K, α) and (K, β) are *cosmetic* if $Y_\alpha(K)$ and $Y_\beta(K)$ are homeomorphic. They are *purely cosmetic* if the homeomorphism is orientation-preserving.

There are many examples of cosmetic surgeries on knots in S^3. For example, if K is an achiral knot in S^3 (i.e, $K = -K$), then $S^3_r(K) \approx -S^3_{-r}(K)$ for any $r \in \mathbb{Q}$. Furthermore, if T is the right-handed trefoil, then

$$S^3_{(18k+9)/(3k+1)}(T) \approx -S^3_{(18k+9)/(3k+2)}(T)$$

for any $k \geq 0$. On the other hand, purely cosmetic surgeries are rare. Gordon conjectured the following, which is known as the *cosmetic surgery conjecture*:

Conjecture 4.101 Let K be a knot in an oriented three-manifold Y such that $Y \setminus N(K)$ is irreducible and not $S^1 \times D^2$. If two surgeries on K are purely cosmetic, then there is an automorphism of $Y \setminus N(K)$ taking one slope to the other.

Again, Heegaard Floer homology has been useful for studying cosmetic surgeries.

Dehn surgery is also used to study hyperbolic three-manifolds. The fundamental result in this direction is the hyperbolic Dehn surgery theorem of Thurston [170][171]:

Theorem 4.102 *Let M be a complete, finite-volume, hyperbolic three-manifold with n cusps. We fix a meridian and longitude for each boundary torus. Then there is a finite set of exceptional slopes for each boundary component such that the result of Dehn filling M along slopes*

$$p_1/q_1, \ldots, p_n/q_n$$

different from the exceptional ones results in a hyperbolic three-manifold. Furthermore, as $p_i^2 + q_i^2 \to \infty$, the manifolds $M(p_1/q_1, \ldots, p_n/q_n)$ converge to M in the geometric topology.

Lackenby and Meyerhoff [92][94] have shown that there are at most 10 exceptional surgeries on a hyperbolic knot complement. In fact, if p/q-surgery is exceptional, then the surgery slope has length at most six and $|q| \leq 8$. If we combine this with Theorem 4.53, we obtain the following result that relates exceptional surgeries with the knot signature:

Proposition 4.103 *If K is a hyperbolic knot in S^3 and p/q is a slope satisfying*

$$|p/q + 2\sigma(K)| > \left(6 + c_1 vol(K) inj(K)^{-3}\right) / |q| \quad or \quad |q| > 8,$$

then the manifold $K(p/q)$ obtained by p/q Dehn surgery along K is hyperbolic.

4.11 Branched Covers

We have seen that every three-manifold can be obtained via surgery on a link in S^3. Taking branched covers of S^3 along links provides another useful method for constructing and studying three-manifolds.

Definition 4.104 Let X and Y be topological spaces. A continuous map $f \colon X \to Y$ is a *branched covering* if there exists a nowhere dense subset $B \subset Y$, called the *branch locus* or *singular set*, such that

$$p|_{X \setminus p^{-1}(B)} \colon X \setminus p^{-1}(B) \to Y \setminus B$$

is a covering map. If Y is connected, the *degree* of the covering is $|p^{-1}(y)|$ for $y \in Y \setminus B$.

First, we look at branched coverings in dimension two. Let S and S' be surfaces. A map $p \colon S' \to S$ is a branched covering if there is a discrete subset $B \subset S$, called the branch locus, such that, if we write $B' = p^{-1}(B)$, then $p|_{S' \setminus B'} \colon S' \setminus B' \to S \setminus B$ is a covering map. In a neighbourhood of each point $b' \in B'$, the map p is conjugate to $z \to z^n$ for an integer $n > 0$ called the branch index of b', which we denote by $\operatorname{ind}(b')$.

Recall that a holomorphic function $f \colon \mathbb{C} \to \mathbb{C}$ such that $f(0) = 0$ has the form $z \mapsto cz^n$ for some $n \in \mathbb{Z}_+$ in suitable coordinate systems about 0 in the source and range. Hence, f is a local homeomorphism if $f'(0) \neq 0$, and is a branch point of some index $n > 1$ if $f'(0) = 0$. Hence, if S and S' are Riemann surfaces (i.e., complex manifolds of complex dimension one) and $p\colon S' \to S$ is a nonconstant holomorphic map, then p is a branched covering. Every closed, connected Riemann surface admits a non-constant holomorphic map to S^2; that is, a meromorphic function.

For a d-fold unbranched covering $p\colon X \to Y$ between finite CW complexes, we have $\chi(X) = d\chi(Y)$. Hurwitz's theorem extends this to branched coverings of surfaces:

Theorem 4.105 *Let S and S' be compact surfaces, and $f\colon S' \to S$ a branched covering of degree d with branch locus B. Then*

$$\chi(S') = d\chi(S) - \sum_{b' \in p^{-1}(B)} (ind(b') - 1).$$

Proof Choose a triangulation of S such that the branch locus B is part of the zero-skeleton. This can be lifted to a triangulation of S'. Each simplex with inside in $S \setminus B$ has d pre-images in S'. For $b \in B$, the number of pre-images drops by $\operatorname{ind}(b') - 1$ for each $b' \in p^{-1}(\{b\})$. The result follows. □

By a classical result of Alexander, every PL n-manifold can be obtained as a branched cover of the n-sphere. Now we turn our attention to dimension three.

Definition 4.106 Let L be a k-component link in a homology three-sphere Y. Then the *n-fold cyclic branched cover* $\Sigma_n(Y, L)$ of Y along L is obtained by taking the n-fold cover of $Y \setminus N(L)$ corresponding to the kernel of the homomorphism

$$\pi_1(Y \setminus N(L)) \to H_1(Y \setminus N(L)) \cong \mathbb{Z}^k \to \mathbb{Z}_n,$$

where the first map is abelianisation, and the second map is

$$(a_1, \ldots, a_k) \mapsto a_1 + \cdots + a_k \mod n.$$

We then fill the boundary with k solid tori, mapping their meridians to the lifts of n times the meridians of each link component.

Proposition 4.107 *If K is a knot in S^3, then*

$$\det(K) = |H_1(\Sigma_2(S^3, K))|.$$

For a proof, see Lickorish [98]; it is an application of the Mayer–Vietoris sequence. In particular, the branched double cover of S^3 along a knot is a *rational homology three-sphere Y*; that is,

$$H_*(Y; \mathbb{Q}) \cong H_*(S^3; \mathbb{Q}).$$

If the knot K is slice with slice disk D in D^4, then we can take the branched double cover of D^4 along D, which is then a rational homology four-ball W with boundary Y. Hence, if we can obstruct Y from bounding a rational homology four-ball, then we can conclude that K is not slice.

Now consider the Brieskorn manifold $M(p,q,r)$ for integers p, q, $r \geq 2$; see Definition 2.66. This is a *homology three-sphere*; that is,

$$H_*(M(p,q,r)) \cong H_*(S^3),$$

if and only if p, q, r are pairwise relatively prime. The manifold $M(2,3,5)$ is known as the *Poincaré homology sphere*, sometimes also denoted by Σ_P. Poincaré originally conjectured that every homology sphere is homeomorphic to S^3, but found this counterexample and reformulated his conjecture to homotopy spheres. It can also be described by gluing the opposite faces of a dodecahedron with a twist.

The manifolds $M(p,q,r)$ can also be described as branched coverings, due to work of Milnor [117]:

Theorem 4.108 *The Brieksorn manifold $M(p,q,r)$ is homeomorphic to the r-fold cyclic branched cover of S^3, branched along the torus link $T_{p,q}$.*

Proof The variety

$$V := \{ (z_1, z_2, z_3) \in \mathbb{C}^3 : z_1^p + z_2^q + z_3^r = 0 \}$$

is smooth except at the origin. Consider the projection

$$\pi\colon V \setminus \{0\} \to \mathbb{C}^2 \setminus \{0\}.$$
$$(z_1, z_2, z_3) \mapsto (z_1, z_2)$$

If $z_1^p + z_2^q \neq 0$, then $|\pi^{-1}(\{(z_1, z_2)\})| = r$, and the cyclic group C_r of rth roots of unity acts on this pre-image by the rule

$$\omega \cdot (z_1, z_2, z_3) = (z_1, z_2, \omega z_3)$$

for $\omega \in C_r$. Hence, the quotient of $V \setminus \{0\}$ by C_r maps homeomorphically onto $\mathbb{C}^2 \setminus \{0\}$, and $V \setminus \{0\}$ is the r-fold cyclic branched cover of $\mathbb{C}^2 \setminus \{0\}$, branched along the algebraic curve $V_{p,q}$ given by $z_1^p + z_2^q = 0$.

Consider the free action of $\mathbb{R}_+$ on $V \setminus \{0\}$ by

$$t \cdot (z_1, z_2, z_3) := (t^{1/p} z_1, t^{1/q} z_2, t^{1/r} z_3)$$

for $t \in \mathbb{R}_+$. Since every $\mathbb{R}_+$-orbit intersects S^3 once transversely, $V \setminus \{0\}$ is canonically diffeomorphic to $\mathbb{R}_+ \times M(p,q,r)$. Note that the actions of $\mathbb{R}_+$ and C_r on $V \setminus \{0\}$ commute.

Similarly, the action of $\mathbb{R}_+$ on $\mathbb{C}^2 \setminus \{0\}$ given by

$$t \cdot (z_1, z_2) := (t^{1/p} z_1, t^{1/q} z_2)$$

is free and induces a canonical diffeomorphism between $\mathbb{C}^2 \setminus \{0\}$ and $\mathbb{R}_+ \times S^3$. As the projection map π is $\mathbb{R}_+$-equivariant, by quotienting out with $\mathbb{R}_+$, it follows that $M(p, q, r)$ is an r-fold cyclic branched cover of S^3, with branch locus $T_{p,q} = V_{p,q} \cap S^3$; see Exercise 4.42. □

Example 4.109 Suppose that K is a two-bridge link in S^3. Then $\Sigma_2(S^3, K)$ is a lens space. Indeed, $S^3_\pm \cap K$ consists of a pair of unknotted arcs, where S^3_+ and S^3_- are the two hemispheres of S^3. The branched double cover of D^3 along a pair of unknotted arcs is a solid torus, which can be seen by cutting D^3 along two half-disks with boundaries on ∂D^3 and on the two arcs, and gluing two copies of the result. So, $\Sigma_2(S^3, K)$ is obtained by gluing two solid tori and is hence a lens space.

The preceding observation underlies the classification of two-bridge links:

Sketch of Proof of Theorem 4.55 It can be shown that the branched double cover of S^3 along $C(a_1, \ldots, a_n)$ is the lens space $L(p, q)$. If two two-bridge knots are isotopic, the corresponding lens spaces are homeomorphic. Hence, the necessity of the condition follows from the classification of lens spaces given in Section 3.7. □

4.12 Two-Knots

We conclude this chapter with a brief discussion of two-knots, which is relevant to four-manifold topology. Recall that a two-knot is an embedding of S^2 into S^4 or $\mathbb{R}^4$. This could be a topologically locally flat or a smooth embedding, but we will focus on the smooth category.

Given a knot K in S^3, we can construct a two-knot using it via an operation called *spinning*, due to Artin [7]. Intuitively, one removes an unknotted subarc of K, obtaining a tangle T in $\mathbb{R}^3_+$ with endpoints in $\mathbb{R}^2$. Then we rotate $\mathbb{R}^3_+$ around $\mathbb{R}^2$ in $\mathbb{R}^4$, and the surface swept out by K in $\mathbb{R}^4$ is called the *Artin spun* of K.

This construction was extended to twist-spinning by Zeeman [179], and to roll-spinning by Fox [35]. Deform-spinning, due to Litherland [101], is a common generalisation of both, which we now describe. Let K be a knot in S^3, and let $\phi \in \mathrm{Diff}(S^3)$ be an automorphism of S^3 that is the identity in a regular neighbourhood $N(K)$ of K. Choose a point $p \in K$, and let v_p be a normal frame in $T_p S^3$. We write $M(\phi)$ for the mapping torus of ϕ. This contains the torus $S^1 \times K$. Since ϕ is the identity along $N(K)$, we can translate ν_p along $S^1 \times \{p\}$ to obtain a normal framing ν of $S^1 \times \{p\}$. If we perform surgery on $(M(\phi), S^1 \times K)$ along $(S^1 \times \{p\})$, we obtain the *deform-spun* two-knot K_ϕ. Clearly, K_{Id} is the Artin spun of K.

Let $X = S^3 \setminus N(K)$ be the knot exterior, and choose a smooth, monotonic function $\varphi \colon \mathbb{R} \to I$ such that $\varphi(t) = 0$ for $t \leq 0$ and $\varphi(t) = 1$ for $t \geq 1$. The automorphism τ of S^3 is supported in a collar $\partial X \times I$ of ∂X, where it is given by the formula

$$\tau(\overline{x}, \overline{\theta}, t) := (\overline{x}, \overline{\theta + \varphi(t)}, t)$$

for $(\overline{x}, \overline{\theta}, t) \in K \times \mathbb{R}/\mathbb{Z} \times I \approx \partial X \times I$. Furthermore, we let

$$\rho(\overline{x}, \overline{\theta}, t) := (\overline{x + \varphi(t)}, \overline{\theta}, t).$$

Then the *k-twist-spin* of K is K_{τ^k} and the *l-roll-spin* is K_{ρ^l}. The *k-twist l-roll spin* of K is defined to be $K_{\tau^k \rho^l}$. Intuitively, using the original description of Artin spinning, the k-twist-spin of K is obtained by rotating the arc around an axis through its endpoints as we rotate $\mathbb{R}^3_+$ around $\mathbb{R}^2$. Roll-spinning is harder to visualise.

Just like for one-knots, one can define Seifert three-manifolds and fibredness for two-knots. The following result is due to Zeeman:

Theorem 4.110 *Let K be a knot in S^3. Then the k-twist-spun K_{τ^k} is fibred for every $k \neq 0$, with closed fibre the k-fold cyclic branched cover of S^3 branched along K, and monodromy the canonical generator of the group of covering transformations.*

Proof The following argument is due to Miller. As above, pick a point $p \in K$. We can identify $S^3 \setminus N(p)$ with D^3, and $K \setminus N(p)$ is a properly embedded arc a in D^3. Then we obtain K_{τ^k} by capping off $(M(\tau^k|_{D^3}), S^1 \times a)$ with $(D^2 \times S^2, D^2 \times \partial a)$, where $\partial a \subset S^2$ and

$$M(\tau^k|_{D^3}) = I \times D^3 /_{(1,w) \sim (0, \tau^k(w))}$$

is the mapping torus of $\tau^k|_{D^3}$.

Choose a circle-valued Morse function $f \colon D^3 \setminus a \to S^1$, and suppose that $n \neq 0$. Then, for each $\theta \in S^1$, we construct a properly embedded three-manifold Y_θ in $M(\tau^k|_{D^3})$ that give a fibration of $M(\tau^k|_{D^3})$. In particular,

$$Y_\theta := \bigcup_{t \in I} \{t\} \times f^{-1}(\overline{\theta + 2\pi k t}),$$

which is diffeomorphic to the $|k|$-fold cyclic cover of $S^3 \setminus N(K)$; that is, $\Sigma_{|k|}(K) \setminus N(K)$; see Figure 4.13.

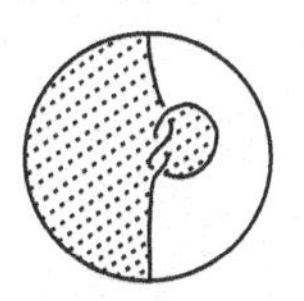 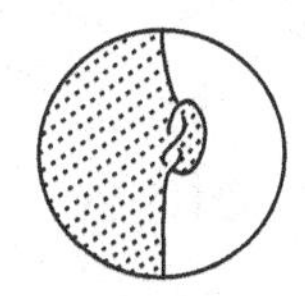 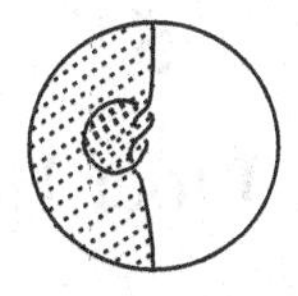 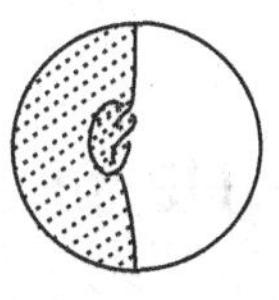 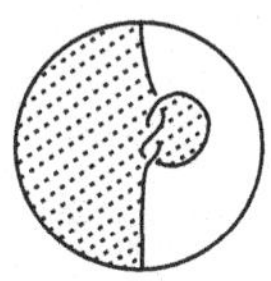

Figure 4.13 The manifold Y_θ for $k = 1$.

Curves in $\partial M(\tau^k|_{D^3})$ that wind n times around a meridian of K and once around the I-factor bound disjoint disks in $D^2 \times S^2$. Hence, we can cap off $Y_\theta \cap \partial M(\tau^k|_{D^3})$ with $D^2 \times I \subset D^2 \times S^2$, where the I-factor corresponds to the arcs $f^{-1}(\theta + 2\pi kt) \cap \partial D^3$, obtaining Seifert manifolds $\hat{Y}_\theta$ for K_{τ^k}. The $\hat{Y}_\theta$ for $\theta \in S^1$ give a fibration of $S^4 \setminus K_{\tau^k}$ over S^1. Furthermore, attaching $D^2 \times I$ to Y_θ with ∂D^2 a meridian of K yields

$$\hat{Y}_\theta = \Sigma_{|k|}(K) \setminus B^3.$$

The monodromy of the fibration is the covering action of $\Sigma_{|k|}(K)$ or its inverse, depending on the sign of k. □

As a special case, we obtain the following surprising corollary:

Corollary 4.111 *If K is a knot in S^3, then the one-twist-spun K_τ is always the trivial two-knot.*

5
Heegaard Floer Homology

Heegaard Floer homology is a package of invariants for oriented three-manifolds [134][137], links [136][140], and oriented four-manifolds [139], defined by Ozsváth and Szabó.

We first give an overview of the necessary background from symplectic topology, followed by the closed three-manifold invariant. Given a closed, oriented three-manifold Y, it is defined in terms of a Heegaard diagram $\mathcal{H} = (\Sigma, \alpha, \beta)$ using symplectic geometry. Invariance under the choice of diagram is achieved using the Reidemeister–Singer theorem for Heegaard diagrams (Theorem 3.35), together with constructing isomorphisms for Heegaard moves: isotopies of the α-curves and β-curves, stabilisations, handle slides, and isotopies of the Heegaard surface in Y. To obtain an invariant of Y that is functorial under diffeomorphisms, one also has to show that the composition of these isomorphisms is independent of the sequence of Heegaard moves [74]. The three-manifold invariant detects the Thurston norm on $H_2(Y)$.

Knot Floer homology is an invariant of knots and links that detects the Seifert genus, fibredness, and gives lower bounds on the four-ball genus. After a brief detour into contact topology and the contact element in Heegaard Floer homology, we outline how to use sutured manifold theory to show that knot Floer homology detects the Seifert genus. We do not discuss the four-manifold invariant here as it is conjecturally equivalent to the Seiberg–Witten invariant, which we review in Chapter 6.

5.1 A Brief Overview of Symplectic Topology

Let V be a $2n$-dimensional $\mathbb{R}$-vector space. A *symplectic form* on V is a non-degenerate, antisymmetric bilinear form; that is, $\omega(v, w) = -\omega(w, v)$. There is a basis $v_1, \ldots, v_n, w_1, \ldots, w_n$ such that $\omega(v_i, v_j) = \omega(w_i, w_j) = 0$

and $\omega(v_i, w_j) = \delta_{ij}$ for every $i, j \in \{1,\ldots,n\}$. For example, $H_1(\Sigma_n;\mathbb{R})$ is a $2n$-dimensional symplectic vector space for an oriented genus n surface Σ_n.

Definition 5.1 A *symplectic manifold* is a pair (M,ω) such that M is a $2n$-manifold, and ω is a closed two-form such that $\omega|_{T_pM}$ is non-degenerate for every $p \in M$.

In particular, $(T_pM, \omega|_{T_pM})$ is a symplectic vector space for every $p \in M$. Non-degeneracy amounts to the condition that $\wedge^n\omega$ is nowhere vanishing and is hence a volume form. So every symplectic manifold has a canonical orientation. For example, any surface with a nowhere vanishing two-form is symplectic.

Unlike Riemannian manifolds, symplectic manifolds all locally look like $\mathbb{R}^{2n}$ with the standard symplectic form

$$\omega_{\text{std}} := \sum_{k=1}^{n} \mathrm{d}x_k \wedge \mathrm{d}y_k,$$

and hence have no local invariants beyond their dimension. So the study of symplectic manifolds is always global.

A submanifold L of the symplectic $2n$-manifold (M,ω) is *Lagrangian* if it is n-dimensional and $\omega|_L \equiv 0$. Note that a symplectic form cannot vanish on a submanifold of dimension larger than n. As an example, every embedded curve on a surface is Lagrangian.

The most useful technique in symplectic topology is the study of so-called pseudoholomorphic curves in the manifold. An *almost complex structure* J on the manifold M is an endomorphism of TM such that

$$J^2 = -\mathrm{Id}_{TM}.$$

In comparison, a *complex n-manifold* is a manifold together with an atlas of charts into $\mathbb{C}^n$ such that the transition maps are holomorphic. Every complex manifold has an almost complex structure induced by multiplication with i in each chart. Not every almost complex structure arises this way, and ones that do are called *integrable*. However, on a surface, every almost complex structure is integrable by the uniformisation theorem.

A submanifold N of M is called *totally real* if $T_pN \cap JT_pN = \{0\}$ for every $p \in N$. For example, $\mathbb{R}^n$ is a totally real submanifold of $\mathbb{C}^n$, hence the name.

The almost complex structure J is *tamed* by the symplectic form ω if

$$\omega(v, Jv) > 0$$

for every non-zero tangent vector $v \in TM$. In other words, ω is positive on every J-complex plane. The space of almost complex structures tamed by ω is always non-empty and contractible. We say that J is *compatible* with ω if it is tamed by it and is J-invariant; that is,

$$\omega(Jv, Jw) = \omega(v, w)$$

for every $p \in M$ and $v, w \in T_pM$. In this case, $\omega(v, Jw)$ is a Riemannian metric on M.

A *J-holomorphic* (or *pseudoholomorphic*) curve on an almost complex manifold (M, J) is a map $u\colon F \to M$, where (F, j) is a Riemann surface, and

$$du(jv) = Jdu(v)$$

for every $v \in TF$. When J is integrable, u is called a complex or holomorphic curve. One can effectively study symplectic manifolds by choosing an almost complex structure J tamed by the symplectic form, and looking at moduli spaces of J-holomorphic curves. In this context, a moduli space is a set of J-holomorphic curves that satisfy some conditions (e.g., lie in a given homotopy or homology class and have prescribed boundary behaviour), endowed with a topology. One of the advantages of considering almost complex structures over complex structures is their flexibility: One can perturb them to make moduli spaces of pseudoholomorphic curves into manifolds. J-holomorphic curves are the solutions of the PDE

$$\bar{\partial} u = 0,$$

and for generic J, solutions are transversely cut out.

The moduli space of pseudoholomorphic curves in a given homology class is not usually compact, but has a nice compactification described by Gromov. The energy of such curves is the integral of the symplectic form over the homology class of the curve, and is hence constant. Given a sequence of pseudoholomorphic curves with energy bounded from above, there exists a subsequence that converges in some sense to a nodal configuration of curves. In the limit, the energy might blow up at certain points, where spheres bubble off, and, in fact, whole trees of such spheres might appear.

Of particular importance to us will be curves with boundary on one or two Lagrangian submanifolds. Given a Lagrangian L in the symplectic manifold (M, ω) with compatible almost complex structure J, one can consider pseudoholomorphic disks

$$u\colon (D^2, S^1) \to (M, L).$$

In this case, Gromov compactness also allows trees of disks bubbling off along $S^1 = \partial D^2$, together with trees of spheres. This happens when the energy blows up along S^1.

When we have two transverse Lagrangians L_0 and L_1 in (M, ω) and intersection points $x, y \in L_0 \cap L_1$, we consider maps $u\colon D^2 \to M$ such that

$S^1 \cap \{\text{Re}(z) \geq 0\}$ maps to L_1 and $S^1 \cap \{\text{Re}(z) \leq 0\}$ maps to L_0; furthermore, $u(-\text{i}) = x$ and $u(\text{i}) = y$. Since $D^2 \setminus \{\text{i}, -\text{i}\}$ is conformally equivalent to $I \times \mathbb{R}$, one can alternatively consider maps $u \colon I \times \mathbb{R} \to M$ such that

$$u(\{k\} \times \mathbb{R}) \subset L_k$$

for $k \in \{0, 1\}$, and $\lim_{z \to -\infty} u(z) = x$, while $\lim_{z \to +\infty} u(z) = y$. A sequence of such pseudoholomorphic strips can bubble in the interior, along the boundary, or could break into the concatenation of two strips that meet along a point of $L_0 \cap L_1$.

Lagrangian intersection Floer homology associates a chain complex

$$(\text{CF}(L_0, L_1), \partial)$$

to a pair of transversely intersecting Lagrangians L_0 and L_1 whose homology we denote by $\text{HF}(L_0, L_1)$. Roughly speaking, $\text{CF}(L_0, L_1)$ is generated by $L_0 \cap L_1$ over some ring, and the boundary map counts rigid pseudoholomorphic curves. However, in general, one might not obtain a chain complex, and there might be issues with smoothness of the moduli spaces, for example. These issues do not arise in the case of Heegaard Floer homology, as we shall soon see.

The motivation behind counting pseudoholomorphic disks is the following. Consider the space $\Omega(L_0, L_1)$ of paths $\gamma \colon I \to M$ with $\gamma(0) \in L_0$ and $\gamma(1) \in L_1$, and let $\ell_0 \in \Omega(L_0, L_1)$ be a basepoint. We write $\Omega_{\ell_0}(L_0, L_1)$ for the path component of ℓ_0. If w is a path in $\Omega(L_0, L_1)$ from ℓ_0 to ℓ, then we can view it as a map

$$w \colon I \times I \to M,$$

such that $w(t, 0) = \ell_0(t)$, $w(t, 1) = \ell(t)$, and $w(i, t) \in L_i$ for $t \in I$ and $i \in \{0, 1\}$. Roughly speaking, Lagrangian Floer homology of L_0 and L_1 computes the homology of a certain cover $\widetilde{\Omega}_{\ell_0}(L_0, L_1)$ of $\Omega_{\ell_0}(L_0, L_1)$ using Morse homology. An element of this cover over $\ell \in \Omega_{\ell_0}(L_0, L_1)$ is an equivalence class $[w]$ of a path w as previously. The role of the Morse function is played by the energy functional

$$\mathcal{A}([w]) := \int_{I \times I} w^* \omega.$$

The issue is that the space $\widetilde{\Omega}_{\ell_0}(L_0, L_1)$ is infinite-dimensional, and hence the index of a critical point is undefined. However, Floer observed that one can define the index difference of two critical points. Critical points of the energy functional are constant paths at the points of $L_0 \cap L_1$, and gradient flow lines between two critical points correspond to pseudoholomorphic curves.

As one might expect, one can also count pseudoholomorphic polygons with boundary along several Lagrangian submanifolds $L_0, \ldots, L_k$. Counting these gives rise to products

$$\text{HF}(L_0, L_1) \otimes \cdots \otimes \text{HF}(L_{n-1}, L_n) \to \text{HF}(L_0, L_n).$$

Such products for triples of Lagrangians give rise to various maps on Heegaard Floer homology.

Lagrangians in a symplectic manifold form the objects of the *Fukaya category*. Morphisms between L_0 and L_1 are the elements of $\mathrm{HF}(L_0, L_1)$, and composition is given by the preceding product for triples. However, these products are not associative, and so one obtains an A_∞-category, where the higher products are given as earlier. For further detail, see Fukaya et al. [40][41] and Seidel [161].

5.2 Heegaard Floer Homology of Three-Manifolds

Let Y be a closed, connected, and oriented three-manifold together with a basepoint z. Given a Heegaard diagram $\mathcal{H} = (\Sigma, \boldsymbol{\alpha}, \boldsymbol{\beta})$ for Y such that $\boldsymbol{\alpha} = \alpha_1 \cup \cdots \cup \alpha_g$ and $\boldsymbol{\beta} = \beta_1 \cup \cdots \cup \beta_g$ intersect transversely and $z \in \Sigma \setminus (\boldsymbol{\alpha} \cup \boldsymbol{\beta})$, one associates a chain complex $\mathrm{CF}(\mathcal{H}, z)$ to it, as follows. First, we choose a complex structure j on Σ. If $g = g(\Sigma)$, we consider the g-fold symmetric product

$$\mathrm{Sym}^g(\Sigma) := \Sigma^g / S_g,$$

which consists of unordered g-tuples of points on Σ.

Exercise 5.2 Use the fundamental theorem of algebra to show that $\mathrm{Sym}^g(\Sigma)$ is a complex manifold.

Let J_t for $t \in \mathbb{R}$ be a generic one-parameter family of almost complex structures on $\mathrm{Sym}^g(\Sigma)$ such that $J_0 = \mathrm{Sym}^g(j)$. For simplicity, we are going to work over the field $\mathbb{F} := \mathbb{F}_2$. The chain complex $\mathrm{CF}(\mathcal{H})$ is the free $\mathbb{F}$-vector space generated by $\mathbb{T}_\alpha \cap \mathbb{T}_\beta$. More concretely, a generator is a g-tuple $\mathbf{x} = (x_1, \ldots, x_g)$ such that $x_i \in \alpha_i \cap \beta_{\sigma(i)}$ for some permutation $\sigma \in S_g$. Hence, the generators can be easily determined from the Heegaard diagram.

The non-combinatorial part of the construction is the boundary map

$$\partial \colon \mathrm{CF}(\mathcal{H}, z) \to \mathrm{CF}(\mathcal{H}, z).$$

We give this via its matrix in the basis $\mathbb{T}_\alpha \cap \mathbb{T}_\beta$. The coefficient of $\mathbf{y} \in \mathbb{T}_\alpha \cap \mathbb{T}_\beta$ in $\partial\mathbf{x}$ counts rigid pseudoholomorphic Whitney disks from $\mathbf{x}$ to $\mathbf{y}$ that avoid the hypersurface $V_z := \{z\} \times \mathrm{Sym}^{g-1}(\Sigma)$. A *topological Whitney disk* is a continuous map $u \colon D^2 \to \mathrm{Sym}^g(\Sigma)$ such that $u(\partial D^2 \cap \{\mathrm{Re}(z) \geq 0\}) \subset \mathbb{T}_\alpha$, $u(\partial D^2 \cap \{\mathrm{Re}(z) \leq 0\}) \subset \mathbb{T}_\beta$, $u(-\mathrm{i})=\mathbf{x}$, and $u(\mathrm{i})=\mathbf{y}$. Let us write $\pi_2(\mathbf{x}, \mathbf{y})$ for the set of homology classes of such topological Whitney disks. For $\phi \in \pi_2(\mathbf{x}, \mathbf{y})$, we denote by $\mathcal{M}(\phi)$ the *moduli space* of pseudoholomorphic representatives of ϕ; that is, maps $u\colon D^2 \to \mathrm{Sym}^g(\Sigma)$ such that

$$du(iv) = J_t(du(v)),$$

where $v \in T_{(t,x)}(I \times \mathbb{R})$ for $t \in I$ and $x \in \mathbb{R}$, and where we conformally identify $D^2 \setminus \{i, -i\}$ with $I \times \mathbb{R}$. If the one-parameter family J_t of almost complex structures is generic, then the moduli space $\mathcal{M}(\phi)$ is a manifold of dimension the Maslov index $\mu(\phi)$ of ϕ, which is a homotopy-theoretic quantity, and, as we shall see, can be read off the Heegaard diagram.

There is an $\mathbb{R}$-action on $\mathcal{M}(\phi)$ obtained by translating $I \times \mathbb{R}$ along the $\mathbb{R}$-direction. We write

$$\widetilde{\mathcal{M}}(\phi) := \mathcal{M}(\phi)/\mathbb{R},$$

and $n_z(\phi)$ is the algebraic intersection number of ϕ and the hypersurface V_z. *Rigid* pseudoholomorphic disks are elements of $\mathcal{M}(\phi)$ for ϕ with $\mu(\phi) = 1$. In this case, the moduli space $\widetilde{\mathcal{M}}(\phi)$ is a compact zero-manifold. The boundary map is given by the formula

$$\partial \mathbf{x} = \sum_{\mathbf{y} \in \mathbb{T}_\alpha \cap \mathbb{T}_\beta} \sum_{\substack{\phi \in \pi_2(\mathbf{x},\mathbf{y}): \\ \mu(\phi)=1,\, n_z(\phi)=0}} \#\widetilde{\mathcal{M}}(\phi) \cdot \mathbf{y},$$

where $\#\widetilde{\mathcal{M}}(\phi)$ is the number of elements of $\widetilde{\mathcal{M}}(\phi)$ modulo 2.

To see that $\partial^2 = 0$, one considers the boundary of one-dimensional moduli spaces, like in the case of Morse homology; see Section 1.5. More specifically, suppose that $\mathbf{x}, \mathbf{z} \in \mathbb{T}_\alpha \cap \mathbb{T}_\beta$ and $\phi \in \pi_2(\mathbf{x}, \mathbf{z})$ has $\mu(\phi) = 2$. The coefficient of $\mathbf{z}$ in $\partial^2 \mathbf{x}$ counts pairs of rigid pseudoholomorphic disks in $\pi_2(\mathbf{x}, \mathbf{y}) \times \pi_2(\mathbf{y}, \mathbf{z})$ for $\mathbf{y} \in \mathbb{T}_\alpha \cap \mathbb{T}_\beta$. These correspond to broken flow lines in Morse homology. The moduli space $\widetilde{\mathcal{M}}(\phi)$ is one-dimensional. It is not compact, but can be compactified with broken flow lines by work of Gromov. Furthermore, the two disks in a broken flow line can be approximated by a sequence of pseudoholomorphic disks; this is called *gluing*. As every one-manifold has an even number of ends, we obtain that the coefficient of $\mathbf{z}$ in $\partial^2 \mathbf{x}$ is zero.

The sum in the definition of the boundary map is not necessarily finite. However, by suitably isotoping $\boldsymbol{\alpha}$, we can obtain a so-called admissible diagram. To define this, we first have to introduce a few concepts. A *domain* in the Heegaard diagram $\mathcal{H}$ is an integer combination of components of $\Sigma \setminus (\boldsymbol{\alpha} \cup \boldsymbol{\beta})$. We denote the $\mathbb{Z}$-module of these by $\mathcal{D}(\mathcal{H})$. They can be thought of as integral two-chains on Σ. We say that the domain $\mathcal{P}$ is *periodic* if $\partial \mathcal{P}$ is an integral combination of components of $\boldsymbol{\alpha}$ and $\boldsymbol{\beta}$.

Definition 5.3 A Heegaard diagram is *admissible* if every periodic domain with coefficient zero at the basepoint has both positive and negative multiplicities.

Given a topological Whitney disk $\phi \in \pi_2(\mathbf{x}, \mathbf{y})$, its domain $\mathcal{D}(\phi)$ is defined to have coefficient $\#(V_w \cap \phi)$ in the component of $\Sigma \setminus (\alpha \cap \beta)$ containing w. We say that a domain $\mathcal{D}$ is *positive*, and write $\mathcal{D} > 0$, if each coefficient of $\mathcal{D}$ is non-negative. If ϕ has a pseudoholomorphic representative, then its domain is positive, since each intersection point of a pseudoholomorphic curve with a complex hypersurface is positive.

We say that a domain $\mathcal{D}$ connects $\mathbf{x}$ to $\mathbf{y}$ if

$$\partial(\partial\mathcal{D} \cap \alpha) = \mathbf{y} - \mathbf{x} \text{ and } \partial(\partial\mathcal{D} \cap \beta) = \mathbf{x} - \mathbf{y}.$$

We denote the set of such domains by $\mathcal{D}(\mathbf{x}, \mathbf{y})$. The map

$$\mathcal{D} \colon \pi_2(\mathbf{x}, \mathbf{y}) \to \mathcal{D}(\mathbf{x}, \mathbf{y})$$

is a bijection. Admissibility ensures that there are only finitely many positive classes in $\mathcal{D}(\mathbf{x}, \mathbf{y})$ of Maslov index one and multiplicity zero at z, and hence the sum defining ∂ is finite.

We are now ready to give a combinatorial formula for the Maslov index, due to Lipshitz [100]. The complex structure determines a Riemannian metric on Σ. We can perturb α and β such that they meet at right angles. If S is a Riemannian surface with k acute right-angled corners and l obtuse right-angled corners, then we define the *Euler measure*

$$e(S) := \chi(S) - k/4 + l/4.$$

Every domain is a linear combination of such surfaces, and we extend e to domains linearly. We let the *point measures* $n_{\mathbf{x}}(\mathcal{D})$ and $n_{\mathbf{y}}(\mathcal{D})$ be the averages of the coefficients of $\mathcal{D}$ at the points of $\mathbf{x}$ and $\mathbf{y}$, respectively. If we write

$$\mu(\mathcal{D}) := n_{\mathbf{x}}(\mathcal{D}) + n_{\mathbf{y}}(\mathcal{D}) + e(\mathcal{D}), \tag{5.1}$$

then $\mu(\phi) = \mu(\mathcal{D}(\phi))$.

The chain complex $\mathrm{CF}(\mathcal{H}, z)$ decomposes as a direct sum along Spin^c structures on M, which correspond to components of the path space $\Omega(\mathbb{T}_\alpha, \mathbb{T}_\beta)$:

Definition 5.4 Let Y be a closed, connected, and oriented three-manifold. We say that the nowhere vanishing vector fields v and w on Y are *homologous* if they are homotopic through nowhere vanishing vector fields in the complement of finitely many balls (i.e., on the two-skeleton of Y). A *Spin^c structure* on Y is a homology class of nowhere vanishing vector fields.

By obstruction theory (see Appendix A.2), $\mathrm{Spin}^c(Y)$ is an affine space over $H^2(Y)$. In other words, given Spin^c structures $\mathfrak{s}$ and $\mathfrak{s}'$, we can define

$$\mathfrak{s} - \mathfrak{s}' \in H^2(Y).$$

If $\mathfrak{s}$ is the homology class of v and $\mathfrak{s}'$ is the homology class of v', then we can view v and v' as sections of the unit tangent bundle STM. Then $\mathfrak{s} - \mathfrak{s}'$ is the first obstruction to homotoping v to v' over the two-skeleton of Y. As TY is trivial for any orientable three-manifold Y, after choosing a trivialisation, we can also view v as a map from Y to S^2, in which case $\mathfrak{s} - \mathfrak{s}'$ is Poincaré dual to $v^{-1}(p) - (v')^{-1}(p)$ for a common regular value p of v and v'. By taking the Euler class of the two-plane field $v^\perp$, we obtain the Chern class $c_1(\mathfrak{s}) \in H^2(Y)$. The map $c_1 \colon \mathrm{Spin}^c(Y) \to H^2(Y)$ satisfies

$$c_1(\mathfrak{s}) - c_1(\mathfrak{s}') = 2(\mathfrak{s} - \mathfrak{s}'),$$

so it is an isomorphism modulo two-torsion.

One can associate a Spin^c structure $\mathfrak{s}(\mathbf{x})$ to every $\mathbf{x} \in \mathbb{T}_\alpha \cap \mathbb{T}_\beta$, as follows. Consider a self-indexing Morse function $f \colon Y \to \mathbb{R}$ with one index zero and one index three critical point, together with a Riemannian metric on Y that induce the Heegaard diagram $\mathcal{H}$. Let $\gamma_\mathbf{x}$ be the union of flow lines of $\mathrm{grad}(f)$ through the points of $\mathbf{x} \cup \{z\}$. The flow line through z connects the index zero and three critical points, and the flow lines through the points of $\mathbf{x}$ give a bijection between the index one and two critical points. The vector field $\mathrm{grad}(f)|_{Y \setminus N(\gamma_\mathbf{x})}$ extends to a nowhere vanishing vector field v on Y by the Poincaré–Hopf index theorem, since each component of $N(\gamma_\mathbf{x})$ contains a positive and a negative singular point of $\mathrm{grad}(f)$. We then set $\mathfrak{s}(\mathbf{x})$ to be the homology class of v.

Given $\mathbf{x}, \mathbf{y} \in \mathbb{T}_\alpha \cap \mathbb{T}_\beta$, the set $\pi_2(\mathbf{x}, \mathbf{y})$ is non-empty if and only if $\mathfrak{s}(\mathbf{x}) = \mathfrak{s}(\mathbf{y})$. Hence, we can define $\mathrm{CF}(\mathcal{H}, z, \mathfrak{s})$ as the subcomplex of $\mathrm{CF}(\mathcal{H}, z)$ generated by those $\mathbf{x} \in \mathbb{T}_\alpha \cap \mathbb{T}_\beta$ for which $\mathfrak{s}(\mathbf{x}) = \mathfrak{s}$. Then

$$\mathrm{CF}(\mathcal{H}, z) = \bigoplus_{\mathfrak{s} \in \mathrm{Spin}^c(Y)} \mathrm{CF}(\mathcal{H}, z, \mathfrak{s}).$$

We can read off $\mathfrak{s}(\mathbf{x}) - \mathfrak{s}(\mathbf{y})$ from the diagram by connecting $\mathbf{x}$ to $\mathbf{y}$ along $\boldsymbol{\alpha}$, and then $\mathbf{y}$ to $\mathbf{x}$ along $\boldsymbol{\beta}$, obtaining a one-cycle whose homology class is Poincaré dual to $\mathfrak{s}(\mathbf{x}) - \mathfrak{s}(\mathbf{y})$. In fact, this is also the homology class of $\gamma_\mathbf{x} - \gamma_\mathbf{y}$.

The homology of $\mathrm{CF}(\mathcal{H}, z)$ only depends on Y, up to isomorphism, and we denote it by $\widehat{\mathrm{HF}}(Y)$. Similarly, we denote the homology of $\mathrm{CF}(\mathcal{H}, z, \mathfrak{s})$ by $\widehat{\mathrm{HF}}(Y, \mathfrak{s})$. To define Heegaard Floer homology functorially, we also need to keep track of the basepoint, and write $\widehat{\mathrm{HF}}(Y, z)$ and $\widehat{\mathrm{HF}}(Y, z, \mathfrak{s})$, respectively.

Example 5.5 Consider the lens space $L(p,q)$. Then $\text{Spin}^c(L(p,q))$ is an affine space over $H^2(L(p,q)) \cong \mathbb{Z}_p$. Consider the genus one Heegaard diagram (T^2,α,β) of $L(p,q)$, where α is a meridian and β is a curve of slope q/p. Then $\mathbb{T}_\alpha = \alpha$ and $\mathbb{T}_\beta = \beta$, and $|\alpha \cap \beta| = p$. Given x, $y \in \alpha \cap \beta$, they lie in the same Spin^c structure if and only if $x = y$. Hence

$$\widehat{\text{HF}}(L(p,q),z,\mathfrak{s}) \cong \mathbb{F}$$

for every $\mathfrak{s} \in \text{Spin}^c(L(p,q))$, so

$$\widehat{\text{HF}}(L(p,q)) \cong \mathbb{F}^p.$$

When $\mathbf{x}$ and $\mathbf{y}$ lie in the same Spin^c structure and there is a domain $\mathcal{D} \in \mathcal{D}(\mathbf{x},\mathbf{y})$ with $\mu(\mathcal{D}) = 1$, in general, it might not be possible to determine the moduli space $\mathcal{M}(\mathcal{D})$. However, this is possible in some cases using the following trick. Let $u\colon D^2 \to \text{Sym}^g(\Sigma)$ be a Whitney disk connecting $\mathbf{x}$ and $\mathbf{y}$, and which is holomorphic with respect to $\text{Sym}^g(j)$ for a complex structure j on Σ. If we pull back the g-fold branched covering

$$\pi\colon \Sigma \times \text{Sym}^{g-1}(\Sigma) \to \text{Sym}^g(\Sigma)$$

along u, we obtain a g-fold branched covering $p\colon S \to D^2$, together with a map $s\colon S \to \Sigma \times \text{Sym}^{g-1}(\Sigma)$. If we compose s with projection π_Σ onto the first factor Σ, we obtain a holomorphic map $h\colon S \to \Sigma$. This is illustrated by the following commutative diagram:

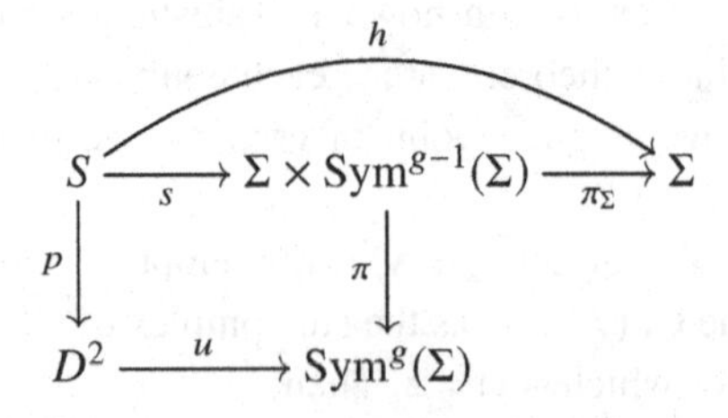

Furthermore, the points of $(\pi \circ s)^{-1}(\mathbf{x})$ and $(\pi \circ s)^{-1}(\mathbf{y})$ divide ∂S into arcs that are alternatingly mapped by h to $\boldsymbol{\alpha}$ and $\boldsymbol{\beta}$.

Conversely, given a g-fold branched covering $p\colon S \to D^2$ and a holomorphic map $h\colon S \to \Sigma$, such that there are marked points along ∂S that are mapped by h to $\mathbf{x}$ and $\mathbf{y}$ and the boundary arcs between them alternatingly to $\boldsymbol{\alpha}$ and $\boldsymbol{\beta}$, then there is a unique holomorphic map $u\colon D^2 \to \text{Sym}^g(\Sigma)$ given by

$$u(x) := h(p^{-1}(x)),$$

where the points of $p^{-1}(x)$ are taken with multiplicity at the branch points.

The above constructions give a bijection between maps $u\colon D^2 \to \mathrm{Sym}^g(\Sigma)$ and tuples (S, p, h). Lipshitz [100] used this to give an alternative construction of Heegaard Floer homology that involves counting pseudoholomorphic curves in $\Sigma \times I \times \mathbb{R}$ instead of $\mathrm{Sym}^g(\Sigma)$.

As an example of the preceding principle, suppose that $\mathbf{x} = \{x_1, \ldots, x_g\}$ and $\mathbf{y} = \{y_1, \ldots, y_g\}$ in $\mathbb{T}_\alpha \cap \mathbb{T}_\beta$ satisfy $x_1 \neq y_1$ and $x_i = y_i$ for $i \in \{2, \ldots, g\}$. Furthermore, there is a bigon between x_1 and y_1 bounded by $\boldsymbol{\alpha}$ and $\boldsymbol{\beta}$. Let $\mathcal{D}$ be the domain that has multiplicity one in the bigon and zero everywhere else. Then $\mu(\mathcal{D}) = 1$ by Lipshitz's formula (5.1) and $\#\widetilde{\mathcal{M}}(\mathcal{D}) = 1$. Indeed, in the unique tuple (S, p, h) in the moduli space, $p\colon S \to D^2$ is the trivial g-fold covering of D^2; that is, S is the disjoint union of g copies of D^2 and p is the identity on each. Furthermore, there are two marked points on each boundary component of S. The map h is constant on $g - 1$ of the components of S with images $x_2 = y_2, \ldots, x_g = y_g$, while the remaining D^2 component is mapped to the bigon connecting x_1 and y_1. By the Riemann mapping theorem, this is unique up to $\mathbb{R}$-translation.

It is only a bit harder to show that $\#\widetilde{\mathcal{M}}(\mathcal{D}) = 1$ also when $x_k = y_k$ for $k \in \{3, \ldots, g\}$ and there is a rectangular region with corners x_1, y_1, x_2, y_2. Again, by the Riemann mapping theorem, there is a conformal map from a disk with four marked points on its boundary to the rectangle if and only if the cross-ratios match. One then studies double covers of D^2 branched along one point and observes that a given cross-ratio can always be uniquely achieved.

Sarkar and Wang [156] showed that one can always isotope $\boldsymbol{\alpha}$ such that every Maslov index one positive domain with multiplicity zero at the basepoint is either a bigon or rectangle, as previously. Hence Heegaard Floer homology is algorithmically computable.

Example 5.6 We now compute $\widehat{\mathrm{HF}}(S^1 \times S^2)$. Consider the Heegaard diagram (T^2, α, β) shown on the left of Figure 5.1, where α and β are two parallel meridians of T^2. Since $\alpha \cap \beta = \emptyset$, one might be tempted to conclude that $\widehat{\mathrm{HF}}(S^1 \times S^2) \cong 0$. However, this is not the case since the diagram is *not* admissible. Indeed, the domain that has multiplicity one in the component of $T^2 \setminus (\alpha \cup \beta)$ that does not contain the basepoint z is periodic, but has no negative coefficient.

If we isotope α as on the right of Figure 5.1, introducing the intersection points $\mathbf{x}, \mathbf{y} \in \alpha \cap \beta$, the diagram becomes admissible. Indeed, every periodic domain that has multiplicity zero at z is a multiple of the periodic domain shown in the figure, with coefficients 1 and -1 at the two bigon components of $T^2 \setminus (\alpha \cup \beta)$, and so has both positive and negative multiplicities. Since both

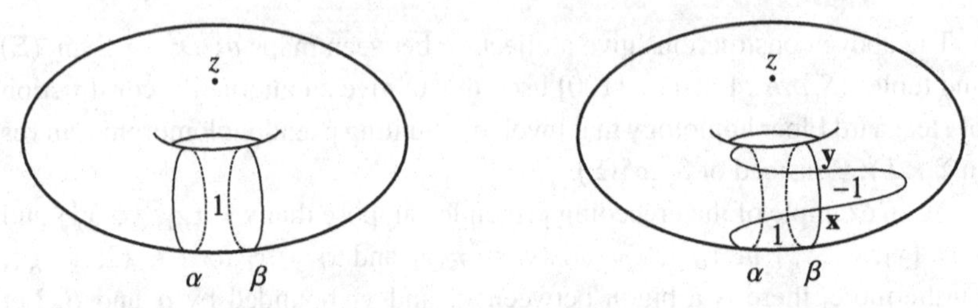

Figure 5.1 The drawing on the left shows a based Heegaard diagram of $S^1 \times S^2$ that is not admissible. We can make this admissible by isotoping α, as shown on the right. The coefficients of some periodic domains in the two diagrams are also shown.

bigons represent a domain from $\mathbf{x}$ to $\mathbf{y}$ and have a single rigid pseudoholomorphic representative up to the $\mathbb{R}$-action, $\partial \mathbf{x} = 0$ as we are working over $\mathbb{F} = \mathbb{F}_2$. There are no other Maslov index one domains, so we conclude that

$$\widehat{\mathrm{HF}}(S^1 \times S^2) \cong \mathbb{F}^2.$$

As a topological application of Heegaard Floer homology, we now outline how it detects the Thurston norm of a closed, connected, and oriented three-manifold. We first introduce the notions involved. Given a compact, connected, and oriented surface S, we let

$$\chi_-(S) := \max\{0, -\chi(S)\}.$$

If S is a compact, oriented surface with components $S_1, \ldots, S_n$, we define its Thurston norm as

$$\Theta(S) := \sum_{i=1}^{n} \chi_-(S_i).$$

For a homology class $\alpha \in H_2(Y)$, the *Thurston seminorm*

$$\Theta(\alpha) := \min\{\, \Theta(S) \colon [S] = \alpha \,\}.$$

There is a twisted version $\underline{\mathrm{HF}}(Y)$ of Heegaard Floer homology, which is a $\mathbb{Z}[H^1(Y)]$-module. This follows from the fact that each component of the path space $\Omega(\mathbb{T}_\alpha, \mathbb{T}_\beta)$ has fundamental group $\mathbb{Z} \oplus H^1(Y)$. Let $\mathbf{x}_0 \in \mathbb{T}_\alpha \cap \mathbb{T}_\beta$ be a fixed intersection point. Then $\pi_2(\mathbf{x}_0, \mathbf{x}_0)$ is the fundamental group of the component of $\Omega(\mathbb{T}_\alpha, \mathbb{T}_\beta)$ containing the constant $\mathbf{x}_0$ path. For $\phi \in \pi_2(\mathbf{x}_0, \mathbf{x}_0)$, the domain $\mathcal{D}(\phi)$ is periodic. We can cap off the boundary of $\mathcal{D}(\phi)$ in Y using the cores of the two-handles attached to Σ along $\boldsymbol{\alpha}$ and $\boldsymbol{\beta}$, giving rise to a two-cycle $H(\mathcal{D}(\phi))$ in Y. We let

$$A(\phi) := \mathrm{PD}\left([H(\mathcal{D}(\phi))]\right) \in H^1(Y).$$

If we fix a topological Whitney disk from $\mathbf{x}_0$ to every point of $\mathbb{T}_\alpha \cap \mathbb{T}_\beta$ in the same Spin^c structure, we can identify $\pi_2(\mathbf{x}, \mathbf{y})$ with $\pi_2(\mathbf{x}_0, \mathbf{x}_0)$. This gives a surjective, additive assignment

$$A\colon \pi_2(\mathbf{x}, \mathbf{y}) \to H^1(Y).$$

The twisted boundary map is defined as

$$\underline{\partial}\mathbf{x} = \sum_{\mathbf{y} \in \mathbb{T}_\alpha \cap \mathbb{T}_\beta} \sum_{\substack{\phi \in \pi_2(\mathbf{x}, \mathbf{y}): \\ \mu(\phi)=1,\, n_z(\phi)=0}} \#\widetilde{\mathcal{M}}(\phi) e^{A(\phi)} \cdot \mathbf{y}.$$

The twisted Heegaard Floer homology $\underline{\widehat{\mathrm{HF}}}(Y)$ also decomposes along Spin^c structures on Y. By the work of Ozsváth and Szabó [136], twisted Heegaard Floer homology detects the Thurston norm of a three-manifold Y in the following sense:

Theorem 5.7 *Let Y be a closed, connected, and oriented three-manifold. For $\alpha \in H_2(Y)$, the Thurston seminorm*

$$\Theta(\alpha) = \max\left\{ |\langle c_1(\mathfrak{s}), \alpha \rangle| : \mathfrak{s} \in Spin^c(Y) \text{ such that } \underline{\widehat{HF}}(Y, \mathfrak{s}) \neq 0 \right\}.$$

5.3 Knot Floer Homology

We have already mentioned in Section 4.9 that one can define concordance homomorphisms using knot Floer homology that give lower bounds on the four-ball genus of a knot. Furthermore, we saw in Section 4.10 that knot Floer homology is useful for studying Dehn surgery problems, such as the Berge conjecture and the cosmetic surgery conjecture.

Given a null-homologous knot K in a closed, connected, and oriented three-manifold Y, knot Floer homology assigns to it a finite-dimensional $\mathbb{F}_2$-vector space $\widehat{\mathrm{HFK}}(Y, K)$. It is a refinement of the three-manifold invariant $\widehat{\mathrm{HF}}(Y)$ and was defined independently by Ozsváth–Szabó [136] and Rasmussen [151].

We will focus only on the case $Y = S^3$, when we write $\widehat{\mathrm{HFK}}(K)$ for $\widehat{\mathrm{HFK}}(S^3, K)$. This group is bigraded; that is,

$$\widehat{\mathrm{HFK}}(K) = \bigoplus_{i,j \in \mathbb{Z}} \widehat{\mathrm{HFK}}_j(K, i),$$

where i is called the *Alexander grading* and j the *homological grading*. If we write

$$h_{i,j} := \dim \widehat{\mathrm{HFK}}_i(K, j),$$

then we can encode knot Floer homology as a two-variable Laurent polynomial

$$p_K(x,y) := \sum_{i,j\in\mathbb{Z}} h_{i,j}x^i y^j.$$

The *graded Euler characteristic* of $\widehat{\mathrm{HFK}}(K)$ is $p_K(x,-1)$, which agrees with the Alexander–Conway polynomial $\Delta_K(x)$. Hence, knot Floer homology is called a *categorification* of the Alexander–Conway polynomial. This terminology is further justified by the fact that (decorated) knot cobordisms functorially induce maps on knot Floer homology.

Knot Floer homology detects the Seifert genus of a knot in the sense that

$$g(K) = \deg(p_K(x,1)). \tag{5.2}$$

This is due to Ozsváth–Szabó [136], and we will outline a simpler proof in Section 5.5. Equation (5.2) strengthens the fact that the degree of the Alexander–Conway polynomial gives a lower bound on the Seifert genus. Further, by work of Ghiggini [46], Ni [129][130], and the author [73], the knot K is fibred if and only if the leading coefficient of $p_K(x,1)$ is ± 1. This extends Proposition 4.68, which states that the Alexander polynomial of a fibred knot is monic.

To define knot Floer homology of a knot K in S^3, we represent K with a doubly-pointed Heegaard diagram $(\Sigma,\boldsymbol{\alpha},\boldsymbol{\beta},w,z)$, as follows. We choose a self-indexing Morse function $f\colon S^3 \to \mathbb{R}$ that has a single index zero critical point m and one index three critical point M, and such that $\mathrm{Crit}(f|_K) = \{m, M\}$. We do this by extending such a smooth function from K to S^3. Then $\Sigma := f^{-1}(3/2)$, we define $\boldsymbol{\alpha}$ and $\boldsymbol{\beta}$ as usual, and the basepoints are $\{w,z\} = K \cap \Sigma$. These are labelled such that the intersection sign of K and Σ is positive at z and negative at w.

Conversely, given a doubly pointed diagram $(\Sigma,\boldsymbol{\alpha},\boldsymbol{\beta},w,z)$, we obtain a knot in S^3 by choosing a compatible Morse function and gradient-like vector field on the three-manifold defined by $(\Sigma,\boldsymbol{\alpha},\boldsymbol{\beta})$ and taking the union of the gradient flow lines through w and z. Equivalently, connect w to z in $\Sigma\setminus\boldsymbol{\alpha}$ and push the resulting arc into the α-handlebody, then connect z to w in $\Sigma\setminus\boldsymbol{\beta}$ and push the resulting arc into the β-handlebody.

The basepoints w and z give rise to an increasing filtration on the chain complex $\mathrm{CF}(\Sigma,\boldsymbol{\alpha},\boldsymbol{\beta},w)$ as follows. There is a unique function $\mathcal{F}\colon \mathbb{T}_\alpha\cap\mathbb{T}_\beta \to \mathbb{Z}$ such that

$$\mathcal{F}(\mathbf{x}) - \mathcal{F}(\mathbf{y}) = n_z(\phi) - n_w(\phi)$$

for any $\phi \in \pi_2(\mathbf{x},\mathbf{y})$, and satisfies the symmetry

$$|\{\mathbf{x}\in\mathbb{T}_\alpha\cap\mathbb{T}_\beta : \mathcal{F}(\mathbf{x}) = i\}| \equiv |\{\mathbf{x}\in\mathbb{T}_\alpha\cap\mathbb{T}_\beta : \mathcal{F}(\mathbf{x}) = -i\}| \mod 2$$

for every $i \in \mathbb{Z}$. Note that the intersection number between a pseudoholomorphic curve and a complex hypersurface is always non-negative. Hence

$$\mathcal{F}(K,i) := \{\mathbf{x} \in \mathbb{T}_\alpha \cap \mathbb{T}_\beta : \mathcal{F}(\mathbf{x}) \leq i\}$$

is a filtration since every $\phi \in \pi_2(\mathbf{x},\mathbf{y})$ with $n_w(\phi) = 0$ and having a pseudoholomorphic representative satisfies $n_z(\phi) \geq 0$, and so $\mathcal{F}(\mathbf{x}) - \mathcal{F}(\mathbf{y}) \geq 0$ if $\mathbf{y}$ appears in $\partial\mathbf{x}$ with non-zero multiplicity. The filtered chain homotopy type of this complex is a knot invariant.

The associated graded complex $\widehat{\mathrm{CFK}}_j(K,i)$ is generated by $\mathbb{T}_\alpha \cap \mathbb{T}_\beta$, and the coefficient of $\mathbf{y}$ in $\partial\mathbf{x}$ is obtained by counting rigid pseudoholomorphic disks from $\mathbf{x}$ to $\mathbf{y}$ disjoint from both V_w and V_z. These have domains with coefficients zero at w and z.

The Alexander grading difference of $\mathbf{x}$ and $\mathbf{y}$ is $2(n_z(\phi) - n_w(\phi))$ for $\phi \in \pi_2(\mathbf{x},\mathbf{y})$. The grading is normalised such that the graded Euler characteristic gives the Alexander–Conway polynomial. The homological grading is pinned down using a similar symmetry property.

Since $(\Sigma, \boldsymbol{\alpha}, \boldsymbol{\beta}, w)$ is a pointed Heegaard diagram of S^3, its Floer homology is $\widehat{\mathrm{HF}}(S^3, w) \cong \mathbb{Z}$. For i small, we have $H_*(\mathcal{F}(K,i)) = 0$, while $H_*(\mathcal{F}(K,i)) \cong \mathbb{Z}$ for i sufficiently large. We define the knot invariant

$$\tau(K) := \min\{i \in \mathbb{Z} \colon H_*(\mathcal{F}(K,i)) \to \widehat{\mathrm{HF}}(S^3) \text{ is non-trivial}\}.$$

The integer $\tau(K)$ is a concordance invariant such that

$$|\tau(K)| \leq g_4(K).$$

If K is alternating, then $2\tau(K) = -\sigma(K)$.

The torus knot $T_{p,q}$ is fibred, and the genus of the fibre is $(p-1)(q-1)/2$. On the other hand, $\tau(T_{p,q}) = (p-1)(q-1)/2$. This implies that

$$g_4(T_{p,q}) = \frac{(p-1)(q-1)}{2},$$

which was conjectured by Milnor; see Proposition 4.43.

Hedden [60] proved that, if $D_+^i(K)$ is the ith iterated zero-twisted Whitehead double of K (see Example 4.48), then $\tau(K) = 0$ if and only if $\tau(D_+^i(K)) = 0$, which supports Conjecture 4.80. Hence, if $\tau(K) \neq 0$, then $D_+^i(K)$ is not smoothly slice for every integer $i > 0$. By Exercise 4.49, every Whitehead double has Alexander polynomial one, and, by Theorem 4.79, every Alexander polynomial one knot is topologically slice. As the trefoil knot $T_{2,3}$ is alternating and $\sigma(T_{2,3}) = -2$, we have $\tau(T_{2,3}) = 1$. Hence, the ith iterated Whitehead double $D_+^i(T_{2,3})$ is topologically slice but not smoothly slice for any i. Such knots underlie the construction of exotic smooth structures on $\mathbb{R}^4$.

Knot Floer homology can be defined completely combinatorially using grid diagrams. The book of Ozsváth, Stipsicz, and Szabó [134] gives a self-contained treatment of this theory. Computing knot Floer homology from grid diagrams is not very efficient. More recently, Ozsváth and Szabó [142] developed an algorithm that can compute knot Floer homology (and hence $g(K)$ and $\tau(K)$) for large knots in seconds. This algorithm is now part of the SnapPy package [27].

5.4 Contact Structures

Heegaard Floer homology has been particularly useful for studying contact three-manifolds. Contact topology can be viewed as the odd-dimensional counterpart of symplectic topology, as contact structures provide natural boundary conditions for symplectic manifolds.

Definition 5.8 A *contact structure* on a $(2n+1)$-manifold M is a nowhere integrable $2n$-plane field; that is, for every $x \in M$, there is no hypersurface in M that is tangent to ξ in a neighbourhood of x. By the Frobenius integrability condition, this amounts to the existence of a local one-form α on M, called a *contact form*, such that $\xi = \ker(\alpha)$ and $\alpha \wedge (\mathrm{d}\alpha)^n$ is nowhere zero.

When both M and ξ are oriented, the form α also exists globally. In this case, we say that ξ is a positive contact structure if $\alpha \wedge (\mathrm{d}\alpha)^n > 0$, and is negative otherwise. Given a contact structure, the contact form is not unique. The restriction $\alpha|_\xi$ is a symplectic form on ξ_x for every $x \in M$.

Two contact structures are said to be *isotopic* if they are isotopic through contact plane fields, and are *contactomorphic* if there is a diffeomorphism taking one to the other.

Example 5.9 The standard contact form on $\mathbb{R}^3$ is $\alpha_{\text{std}} := \mathrm{d}z - y\mathrm{d}x$. We check that

$$\alpha \wedge \mathrm{d}\alpha = (\mathrm{d}z - y\mathrm{d}x) \wedge (\mathrm{d}x \wedge \mathrm{d}y) = \mathrm{d}x \wedge \mathrm{d}y \wedge \mathrm{d}z.$$

Every contact three-manifold is locally contactomorphic to $(\mathbb{R}^3, \xi_{\text{std}})$. Hence, like symplectic manifolds, contact three-manifolds also have no local invariants.

Definition 5.10 A link L in a contact three-manifold (Y, ξ) is called *Legendrian* if it is tangent to ξ. We say that L is *transverse* if it is transverse to ξ.

Contact three-manifolds can be obtained by doing so-called contact surgery along Legendrian links; see Ozbagci–Stipsicz [133]. Every Legendrian link can be perturbed into a transverse link.

Definition 5.11 The contact three-manifold (Y,ξ) is *overtwisted* if there is a disk D embedded in Y such that ∂D is Legendrian, and is *tight* otherwise.

By work of Eliashberg [31], overtwisted contact structures on a closed, connected, and oriented three-manifold Y satisfy an h-principle, and hence there is a unique overtwisted contact structure up to isotopy in every homotopy class of two-plane fields. By obstruction theory (see Appendix A.2), the first obstruction to homotoping two oriented two-plane fields is an element of $H^2(Y;\pi_2(S^2)) = H^2(Y)$, and it is the Spin^c structure represented by $\xi^\perp$. The second obstruction is an element of $H^3(Y;\pi_3(S^2)) \cong \mathbb{Z}$. Consequently, most attention has been devoted to the study of tight contact structures. Classifying these on a given three-manifold is usually difficult. Eliashberg showed that S^3 admits a unique tight contact structure up to isotopy, namely ξ_{std}.

Given a contact three-manifold (Y,ξ), Ozsváth and Szabó [139] associated to it an invariant

$$c(\xi) \in \widehat{\mathrm{HF}}(-Y).$$

This vanishes when ξ is overtwisted, though there are tight contact structures for which it is also zero. For example, $c(\xi) = 0$ whenever ξ has positive Giroux torsion, which is $T^2 \times I$ embedded in Y with a specific contact structure. However, $c(\xi) \neq 0$ if ξ is Stein fillable.

One can encode a contact structure on a three-manifold using open book decompositions.

Definition 5.12 Let (Y,ξ) be a contact three-manifold. An *open book decomposition* of (Y,ξ) consists of a fibred link $B \subset Y$ called the *binding*, together with a fibration of $\phi\colon Y \setminus B \to S^1$ such that

(i) B is positively transverse to ξ,
(ii) there is a contact form α for ξ such that $\alpha|_{\phi^{-1}(\theta)}$ is an area form for every $\theta \in S^1$.

The fibres of ϕ are called *pages* of the open book decomposition.

By the *Giroux correspondence* [47], there is a bijection between contact three-manifolds and open book decompositions up to an operation called positive stabilisation.

An *abstract open book decomposition* consists of a compact surface S with boundary and an automorphism $\varphi\colon S \to S$ that is the identity on ∂S. This defines a three-manifold Y by taking $I \times S$, and identifying $(1, x)$ with $(0, \varphi(x))$ for every $x \in S$, and (t, x) with (t', x) for every $t, t' \in I$ and $x \in \partial S$. There is a unique contact structure ξ on the resulting three-manifold up to isotopy that satisfies the conditions of Definition 5.12.

Honda, Kazez, and Matić [67] have given an alternative definition of the contact class in Heegaard Floer homology using open book decompositions: Let (Y, ξ) be a contact three-manifold with open book decomposition

$$\phi\colon Y \setminus B \to S^1.$$

We obtain a Heegaard surface Σ by smoothing $\phi^{-1}(1) \cup \phi^{-1}(-1)$. Then we take a collection of properly embedded and pairwise disjoint arcs $a_1, \ldots, a_n$ on the page $S := \phi^{-1}(1)$ that form a basis of $H_1(S, \partial S)$. The arcs $b_1, \ldots, b_n$ are small isotopic translates of $a_1, \ldots, a_n$ such that a_i and b_i intersect in a single positive intersection point θ_i, and ∂a_i is moved along ∂S in the positive direction. We write a_i' for the curve on $\phi^{-1}(-1)$ corresponding to a_i, and b_i' for the result of applying the monodromy of the fibration ϕ to a_i'. If we set $\alpha_i := a_i \cup a_i'$ and $\beta_i := b_i \cup b_i'$, then $(\Sigma, \{\beta_1, \ldots, \beta_n\}, \{\alpha_1, \ldots, \alpha_n\})$ is a Heegaard diagram of $-Y$, and the intersection point

$$\boldsymbol{\theta} := \{\theta_1, \ldots, \theta_n\} \in \mathbb{T}_\alpha \cap \mathbb{T}_\beta$$

represents the contact class $c(\xi) \in \mathrm{HF}(-Y)$. In order for $\boldsymbol{\theta}$ to be a cycle, we place the basepoint z on ∂S.

The original proof of Ozsváth and Szabó that knot Floer homology detects the Seifert genus (Equation (5.2)) used the contact invariant. A simpler proof can be given using sutured Floer homology, which we discuss next.

5.5 Sutured Floer Homology

We conclude our overview of Heegaard Floer homology by outlining the proof of the claim in Section 5.3 that knot Floer homology detects the Seifert genus. The tool that we will use is sutured Floer homology [72][73].

We first encountered sutured manifolds in Section 4.7 when we gave a method for deciding whether a knot is fibred using product disk decompositions. In fact, one can decompose sutured manifolds along arbitrary properly embedded surfaces:

Definition 5.13 Let (M, γ) be a sutured manifold and S a compact, oriented, properly embedded surface in M. Then the *sutured manifold decomposition*

$$(M,\gamma) \overset{S}{\rightsquigarrow} (M',\gamma')$$

consists of cutting M along the surface S, resulting in the manifold M'. We add the positive side of S to $R_+(\gamma)$ and the negative side to $R_-(\gamma)$. The suture γ' on $\partial M'$ is where the new R_+ meets the new R_-.

A *sutured manifold hierarchy* is a sequence of sutured manifold decompositions that results in a collection of three-balls with a single suture on each.

A sutured manifold hierarchy is a refinement of a Haken hierarchy. We showed in Proposition 3.13 that every three-manifold M with boundary and $b_1(M) > 0$ is Haken. Recall that we introduced the Thurston norm at the end of Section 5.2.

Definition 5.14 We say that the sutured manifold (M,γ) is *taut* if M is irreducible and $R(\gamma)$ is incompressible and Thurston norm minimising in its relative homology class in $H_2(M,\gamma)$.

Gabai [44] showed the following deep result:

Theorem 5.15 *The sutured manifold (M,γ) is taut if and only if it admits a sutured manifold hierarchy.*

The key to the proof of the forward direction is to define a complexity of sutured manifolds that decreases under certain sutured manifold decompositions that Gabai called *well-groomed*, and showing that one can always find such a decomposition that results in a taut sutured manifold unless the manifold is a collection of three-balls.

Definition 5.16 The sutured manifold (M,γ) is *balanced* if

$$\chi(R_+(\gamma)) = \chi(R_-(\gamma)),$$

the manifold M has no closed components, and each component of ∂M has at least one suture.

Sutured Floer homology $SFH(M,\gamma)$ for a balanced sutured manifold (M,γ) is an extension of the hat version of Heegaard Floer homology to three-manifolds with boundary [72]. To define it, one has to extend Heegaard diagrams to sutured manifolds first.

Definition 5.17 A *sutured diagram* is a tuple $(\Sigma, \boldsymbol{\alpha}, \boldsymbol{\beta}, \mathbf{z})$, where Σ is a closed, connected, and oriented surface, $\boldsymbol{\alpha}$ and $\boldsymbol{\beta}$ are one-dimensional submanifolds of Σ with the same number of components, $\mathbf{z} \subset \Sigma \setminus (\boldsymbol{\alpha} \cup \boldsymbol{\beta})$ is finite, and the components of $\boldsymbol{\alpha}$ and $\boldsymbol{\beta}$ are linearly independent in $H_1(\Sigma_0)$ for $\Sigma_0 := \Sigma \setminus N(\mathbf{z})$.

Note that we do not require $g(\Sigma)$ to agree with $|\boldsymbol{\alpha}| = |\boldsymbol{\beta}|$. A sutured diagram defines a balanced sutured manifold (M, γ) as follows. The three-manifold M is obtained by attaching three-dimensional two-handles to $\Sigma_0 \times I$ along $\boldsymbol{\alpha} \times \{0\}$ and $\boldsymbol{\beta} \times \{1\}$. The sutures are defined to be $\gamma := \partial \Sigma_0 \times I$. One can show analogously to Proposition 3.28 that every balanced sutured manifold admits such a diagram.

Exercise 5.18 Show that the sutured manifold defined by a sutured diagram is indeed balanced.

Given a balanced sutured manifold (M, γ), we define $SFH(M, \gamma)$ by considering the tori $\mathbb{T}_\alpha = \alpha_1 \times \cdots \times \alpha_d$ and $\mathbb{T}_\beta = \beta_1 \times \ldots \beta_d$ in $\mathrm{Sym}^d(\Sigma)$, where $\alpha_1, \ldots, \alpha_d$ are the components of $\boldsymbol{\alpha}$ and $\beta_1, \ldots, \beta_d$ are the components of $\boldsymbol{\beta}$, and taking the $\mathbb{F}_2$-vector space generated by $\mathbb{T}_\alpha \cap \mathbb{T}_\beta$. The boundary map counts rigid pseudoholomorphic disks that avoid V_z for every $z \in \mathbf{z}$.

Since we can always represent a product sutured manifold $(R \times I, \partial R \times I)$ for a compact, connected, and oriented surface R having $\partial R \neq \emptyset$ with the diagram $(R, \emptyset, \emptyset, \emptyset)$, we have that

$$SFH(R \times I, \partial R \times I) \cong \mathbb{F}_2.$$

The key results that we need are the following [72]. The first one describes what happens to sutured Floer homology under a sutured manifold decomposition. To state it, we need a definition first.

Definition 5.19 Given a compact, oriented surface R with no closed components, we say that a curve $C \subset R$ is *boundary-coherent* if either $[C] \neq 0$ in $H_1(R)$, or if $[C] = 0$ and C is oriented as the boundary of its interior.

We say that a decomposing surface S in (M, γ) is *nice* if it has no closed component, and for every component V of $R(\gamma)$, the set of closed components of $S \cap V$ consists of parallel-oriented boundary-coherent simple closed curves.

In particular, every well-groomed sutured manifold decomposition is nice.

Theorem 5.20 *Let* $(M, \gamma) \overset{S}{\rightsquigarrow} (M', \gamma')$ *be a sutured manifold decomposition such that* (M, γ) *and* (M', γ') *are balanced and* S *is nice. Then*

$$\dim SFH(M', \gamma') \leq \dim SFH(M, \gamma).$$

Sketch of Proof One first finds a sutured diagram $(\Sigma, \boldsymbol{\alpha}, \boldsymbol{\beta}, \mathbf{z})$ for (M, γ) where $\Sigma_0 := \Sigma \setminus N(\mathbf{z})$ contains a subsurface P with polygonal boundary whose corners lie on $\partial\Sigma_0$. Furthermore, $\partial P = A \cup B$, where the edges of ∂P alternately lie in A or B, and $\boldsymbol{\alpha} \cap B = \emptyset$ and $\boldsymbol{\beta} \cap A = \emptyset$, and S is obtained by smoothing the corners of

$$P \cup (A \times [1/2, 1]) \cup (B \times [0, 1/2]).$$

One can obtain such a diagram by extending a suitable Morse function from S to M. Then one isotopes $\boldsymbol{\alpha}$ and $\boldsymbol{\beta}$ using the Sarkar–Wang algorithm [156] such that every Maslov index one positive domain becomes either a bigon or a rectangle.

One obtains a diagram $(\Sigma', \boldsymbol{\alpha}', \boldsymbol{\beta}', \mathbf{z}')$ for (M', γ') by taking the trivial double cover of Σ_0 over P (the cover is one-to-one over $\Sigma_0 \setminus P$) and lifting $\boldsymbol{\alpha}$ and $\boldsymbol{\beta}$. Clearly,

$$\mathbb{T}_{\alpha'} \cap \mathbb{T}_{\beta'} \subset \mathbb{T}_{\alpha} \cap \mathbb{T}_{\beta}.$$

Furthermore, every Maslov index one positive domain in $\Sigma' \setminus N(\mathbf{z}')$ is a lift of such a domain in the original diagram, so $\mathrm{CF}(\Sigma', \boldsymbol{\alpha}', \boldsymbol{\beta}', \mathbf{z}')$ is a subcomplex of $\mathrm{CF}(\Sigma, \boldsymbol{\alpha}, \boldsymbol{\beta}, \mathbf{z})$. □

The second result relates sutured Floer homology to knot Floer homology:

Proposition 5.21 *Let K be a knot in S^3 and R a Seifert surface for K of genus g. If $S^3(R)$ denotes the sutured manifold complementary to R (see Section 4.7), then*

$$\widehat{HFK}(K, g) \cong SFH(S^3(R)).$$

Sketch of Proof This follows from a refinement of Theorem 5.20, where one identifies the Spin^c structures of the generators that survive under a sutured manifold decomposition. These are exactly the ones that can be represented by vector fields that never point in the direction of the negative normal of the decomposing surface, and are called *outer*.

If one considers the sutured manifold $S^3(K) = (M, \gamma)$, where M is the knot exterior and γ consists of two oppositely oriented meridians of K, then it is a tautology that

$$SFH(S^3(K)) \cong \widehat{\mathrm{HFK}}(K).$$

Decomposing $S^3(K)$ along the Seifert surface R gives rise to $S^3(R)$. The result follows once one shows that there is exactly one Spin^c structure $\mathfrak{s} \in \mathrm{Spin}^c(S^3(K))$ that is outer with respect to R and

$$SFH(S^3(K), \mathfrak{s}) \cong \widehat{\mathrm{HFK}}(K, g).$$

□

Finally, we have the following vanishing result:

Proposition 5.22 *Let (M,γ) be a balanced sutured manifold such that $R(\gamma)$ is* not *Thurston norm minimising. Then $SFH(M,\gamma) = 0$.*

This is shown by constructing a diagram for (M,γ) where $\mathbb{T}_\alpha \cap \mathbb{T}_\beta = \emptyset$.

Combining the preceding results, we obtain the following, which is equivalent to Equation (5.2):

Theorem 5.23 *Let K be a knot in S^3. Then*

$$\dim \widehat{HFK}(K, g(K)) \geq 1$$

and $\dim \widehat{HFK}(K,g) = 0$ *for $g > g(K)$.*

Proof Let R be a Seifert surface for K of genus g, and consider the complementary sutured manifold $S^3(R)$. By Proposition 5.21,

$$\widehat{\mathrm{HFK}}(K,g) \cong SFH(S^3(R)).$$

If $g = g(K)$, then $S^3(R)$ is taut; hence, by Theorem 5.15, it admits a sutured manifold hierarchy

$$S^3(R) \overset{S_1}{\rightsquigarrow} (M_1,\gamma_1) \overset{S_2}{\rightsquigarrow} \cdots \overset{S_n}{\rightsquigarrow} (M_n,\gamma_n),$$

where (M_n,γ_n) is a disjoint union of balls with a single suture on each, and is hence a product. In fact, we can arrange that $S_1,\ldots,S_n$ are all well-groomed and hence nice. So, by Theorem 5.20, we have that

$$\dim SFH(S^3(R)) \geq \dim SFH(M_1,\gamma_1) \geq \cdots \geq \dim SFH(M_n,\gamma_n) = 1.$$

If $g > g(K)$, then $R(S^3(R))$ is not Thurston norm minimising; hence

$$SFH(S^3(R)) = 0$$

by Proposition 5.22. □

6
Four-Manifolds

Four-manifold theory has a completely different flavour than three-manifold topology. This is the first dimension where the topological and the smooth classification differ. Furthermore, since the Whitney trick and hence the h-cobordism theorem and surgery theory fail in dimension four, there are numerous exotic phenomena not present in higher dimensions.

By Theorem 1.13, every finitely presented group appears as the fundamental group of a closed, smooth four-manifold. Furthermore, there is no algorithm to decide whether a finitely presented group is trivial. Hence, unless one restricts the fundamental group, there is no hope of obtaining a classification. Much of four-manifold topology focuses on simply connected manifolds.

By the work of Freedman, the classification of simply connected *topological* four-manifolds is governed by algebraic topology; namely, the intersection form on the second cohomology group.

The classification of *smooth* four-manifolds relies on invariants that can distinguish homeomorphic but non-diffeomorphic manifolds. One of the main tools for constructing and manipulating four-manifolds is Kirby calculus, which allows us to visualise a smooth four-manifold as a framed link in the three-sphere. Furthermore, it gives a set of moves such that two four-manifolds are diffeomorphic if and only if they are related by a sequence of such moves. Some constructions in four-manifold topology originate from algebraic and symplectic geometry, such as blow-ups and fibre sums.

The two main smooth four-manifold invariants are the Donaldson and the Seiberg–Witten invariants. These come from gauge theory. Roughly speaking, one chooses a Riemannian metric on the manifold, writes down some nonlinear PDEs, and counts the solutions. Then one shows that this count is independent of the chosen metric. Heegaard Floer theory is a more combinatorial version of Seiberg–Witten theory. As an application, we conclude this chapter by reviewing a celebrated result of Fintushel and Stern on how to use an operation called knot surgery to construct infinitely many smooth structures on certain four-manifolds, including the K3 surface.

6.1 Kirby Calculus

Kirby calculus gives us a way to represent and manipulate four-manifolds using framed links in the three-sphere. If the four-manifold has boundary, it can also be used to study the boundary three-manifold. More generally, one can even use it to study four-dimensional cobordisms.

Let X be a smooth, connected, and oriented four-manifold (not necessarily closed). Then it admits a handle decomposition with a single zero-handle. The zero-handle can be identified with D^4, which has boundary S^3. Each one-handle is attached along a framed zero-sphere, which one can represent by a pair of two-spheres, with their boundaries identified by an orientation-reversing diffeomorphism–usually a reflection.

Alternatively, one can visualise a one-handle h^1 as an unknotted circle with a dot on it: If we attach a two-handle h^2 cancelling h^1, we can identify the resulting manifold with D^4. The belt sphere of h^2 is an unknotted circle in S^3, which we mark with a dot. If D is a two-disk with boundary the dotted circle, we can push it into D^4 to obtain the cocore of h^2, and removing a neighbourhood of it amounts to removing h^2, leaving us with h^1.

After attaching all the one-handles, we obtain the boundary connected sum of a number of copies of $S^1 \times D^3$, whose boundary is $\#_n(S^1 \times S^2)$. The two-handles are attached along a framed link L in $\#_n(S^1 \times S^2)$. If we denote the one-handles with pairs of two-spheres, a component of L is either a simple closed curve disjoint from the one-handles, or a collection of arcs in S^3 with endpoints on the two-spheres, which are paired up using the reflections. In dotted circle notation, we can visualise L as a usual link in S^3. The framing can be encoded by a parallel copy F of L. When there are no one-handles, we can encode the framing by an integer on each component of L, which gives the linking number between the corresponding components of L and F. It is unknown whether every closed, simply connected four-manifold admits a handle decomposition without one-handles.

After the two-handles are attached, we have a four-manifold X_2 with boundary a three-manifold Y. Then the one-handles with the framed link can be viewed as a diagram of X_2, or as a diagram of the boundary three-manifold Y. If X is closed, we further have to attach some three-handles and a single four-handle. For this to be possible, Y has to be a connected sum of copies of $S^1 \times S^2$. If this is the case, by the following result of Laudenbach and Poénaru, this uniquely determines X, up to diffeomorphism:

Proposition 6.1 *Every automorphism of $\#_k(S^1 \times S^2)$ extends to the boundary connected sum $\natural_k(S^1 \times D^3)$.*

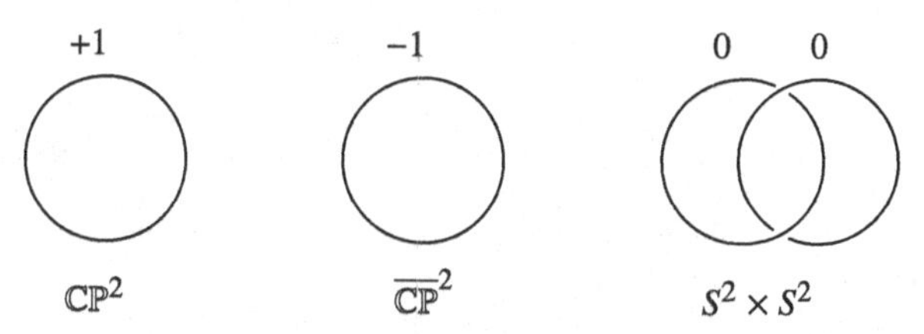

Figure 6.1 Kirby diagrams for $\mathbb{CP}^2$, $\overline{\mathbb{CP}}^2$, and $S^2 \times S^2$.

In other words, there is an essentially unique way to attach the three-handles and four-handles. Hence we do not need to encode these in the diagram.

Example 6.2 We now give examples of Kirby diagrams of some simply connected four-manifolds; see Figure 6.1. One can obtain diagrams for connected sums of these by taking their disjoint union.

The complex projective plane $\mathbb{CP}^2$ with the complex orientation is represented by the unknot with framing $+1$. Indeed, if we remove a ball–thought of as a four-handle–from $\mathbb{CP}^2$, we are left with the total space of the tautological line bundle over $\mathbb{CP}^1$. This has Euler class $+1$, as two complex lines in $\mathbb{CP}^2$ intersect once positively. It can be obtained by gluing two copies of $D^2 \times D^2$ along $D^2 \times S^1$ with framing $+1$. We identify one copy of $D^2 \times D^2$ with the zero-handle B^4. Then the other is a two-handle attached to B^4 along the unknot $\{0\} \times S^1$ with framing $+1$. We denote by $\overline{\mathbb{CP}}^2$ the complex projective plane with its orientation reversed. It can be represented by an unknot with framing -1.

Next, we describe a Kirby diagram for $S^2 \times S^2$. We write S^2 as the union of the hemispheres S^2_+ and S^2_-. Then

$$S^2 \times S^2 = (S^2_- \times S^2_-) \cup (S^2_+ \times S^2_-) \cup (S^2_- \times S^2_+) \cup (S^2_+ \times S^2_+).$$

We view $S^2_- \times S^2_-$ as a zero-handle, $S^2_+ \times S^2_-$ and $S^2_- \times S^2_+$ as two-handles, and $S^2_+ \times S^2_+$ as a four-handle. The two-handles are attached along $\{0\} \times \partial S^2_-$ and $\partial S^2_- \times \{0\}$, respectively, which form a Hopf link. Both components of the Hopf link have framing zero, since the self-intersections of both $S^2 \times \{p\}$ and $\{p\} \times S^2$ are zero for any $p \in S^2$.

By the Lickorish–Wallace theorem (Theorem 4.96), every closed, connected, and oriented three-manifold Y admits a Kirby diagram without one-handles. Let L be the framed link of this Kirby diagram, and we denote by M_L the corresponding four-manifold with $\partial M_L = Y$. The boundary is unchanged under the following two *Kirby moves*:

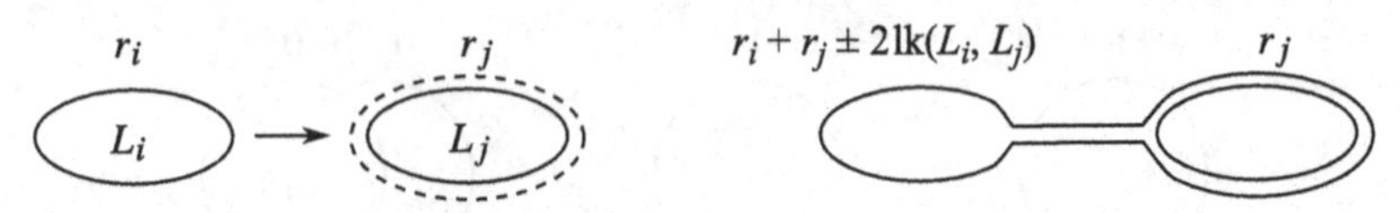

Figure 6.2 Handle sliding the link component L_i with framing r_i over the component L_j with framing r_j.

(i) **Blow-up:** We add or remove an unknot with framing ± 1 that is unlinked from L. If L' is obtained by adding a $+1$-framed unknot to L, then $M_{L'} \approx M_L \# \mathbb{CP}^2$. Similarly, $M_{L'} \approx M_L \# \overline{\mathbb{CP}}^2$ if L' is obtained from L by adding a (-1)-framed unknot.

(ii) **Handle slide:** We do a handle slide among the two-handles. If L_i and L_j are the two components corresponding to the handles involved, with framings F_i and F_j, respectively, then we choose a band connecting L_i and F_j (i.e., an embedding $b\colon I \times I \hookrightarrow M_L$ such that $b(\{0\} \times I) \subset L_i$ and $b(\{1\} \times I) \subset F_j$, while the rest of $b(I \times I)$ is disjoint from L and F), and we replace L_i with $L_i' := L_i \#_b F_j$. For $k \neq i$, we set $L_k' := L_k$ and $F_k' := F_k$. The framing of L_i' is $r_i + r_j \pm 2\text{lk}(L_i, L_j)$, depending on whether the band is orientation-preserving, where $r_k = \text{lk}(L_k, F_k)$. For an illustration, see Figure 6.2.

While blow-up changes the underlying four-manifold M_L, handle slide does not.

There are two S^2-bundles over S^2. Indeed, since D^2 is contractible, the bundle is trivial over the hemispheres S^2_+ and S^2_-, and the bundle is defined by the gluing map along the equator. This is given by an element of $\pi_1(\text{SO}(3)) \cong \mathbb{Z}_2$. We denote the non-trivial S^2-bundle over S^2 by $S^2 \tilde{\times} S^2$.

Exercise 6.3 Using Kirby calculus, show that

$$(S^2 \times S^2) \# \mathbb{CP}^2 \approx \mathbb{CP}^2 \# \mathbb{CP}^2 \# \overline{\mathbb{CP}}^2.$$

Furthermore, $S^2 \tilde{\times} S^2 \approx \mathbb{CP}^2 \# \overline{\mathbb{CP}}^2$.

The real power of Kirby calculus is due to the following result of Kirby [84]:

Theorem 6.4 *Two Kirby diagrams without one-handles represent diffeomorphic three-manifolds if and only if they are related by a finite sequence of Kirby moves* (i) *and* (ii).

Proof It is clear that three-manifolds defined by diagrams related by a sequence of Kirby moves are diffeomorphic.

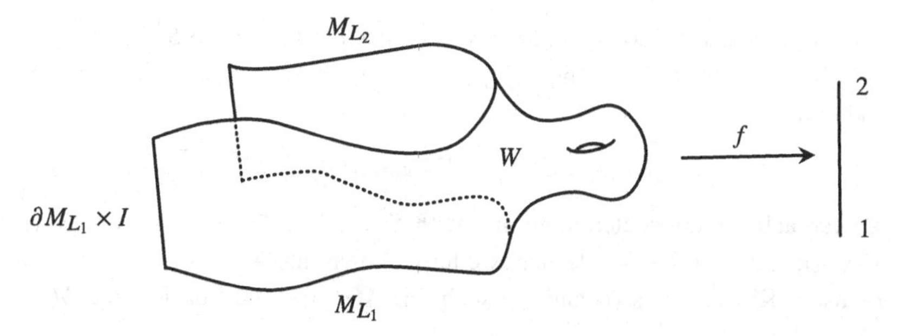

Figure 6.3 The Morse function f on the cobordism W from M_{L_1} to M_{L_2}.

Conversely, suppose that the Kirby diagrams given by the framed links L_1 and L_2 represent diffeomorphic three-manifolds $Y_1 = \partial M_{L_1}$ and $Y_2 = \partial M_{L_2}$. We show that L_1 and L_2 are related by a sequence of Kirby moves.

Step 1. We claim that M_{L_1} and M_{L_2} become diffeomorphic after adding a number of copies of $\mathbb{CP}^2$ and $\overline{\mathbb{CP}}^2$, which can be achieved by Kirby move (i). This is essentially a result of Wall. Consider the four-manifold

$$X := M_{L_1} \cup (\partial M_{L_1} \times [1,2]) \cup -M_{L_2},$$

where we identify ∂M_{L_1} with $\partial M_{L_1} \times \{1\}$ via $x \mapsto (x,1)$ and ∂M_{L_2} with $\partial M_{L_1} \times \{2\}$ via $x \mapsto (d(x),2)$ for a diffeomorphism $d\colon \partial M_{L_2} \to \partial M_{L_1}$. By move (i) applied to L_1, we can add copies of $\mathbb{CP}^2$ and $\overline{\mathbb{CP}}^2$ to M_{L_1} such that $\sigma(X) = 0$, and so

$$[X] = 0 \in \Omega_4^{\mathrm{SO}};$$

see Section 1.7. Then there exists a connected, oriented five-manifold W such that $\partial W = X$; see Figure 6.3.

Choose a Morse function $f\colon W \to [1,2]$ with no index zero and five critical points such that $f^{-1}(i) = M_{L_i}$ for $i \in \{1,2\}$ and $f|_{\partial M_1 \times [1,2]}$ is projection onto $[1,2]$. Suppose that f has a single critical point of index one in $f^{-1}([1,1+\varepsilon])$ for some $\varepsilon > 0$. Then $f^{-1}(1+\varepsilon) = M_{L_1} \# (S^1 \times S^3)$. We can replace the one-handle with a three-handle, which does not change $f^{-1}(1+\varepsilon)$, since

$$\partial D^2 \times S^3 = S^1 \times S^3 = S^1 \times \partial D^4$$

(this is an instance of surgery). Similarly, we can replace index four critical points of f with index two critical points. We can rearrange the critical points such that all index two critical points lie in $f^{-1}((1,3/2))$ and all index three critical points lie in $f^{-1}((3/2,2))$. Since M_{L_1} is simply connected, we can contract the attaching circles of the two-handles of W, so $f^{-1}(3/2)$ can be obtained from M_{L_1} by adding copies of $S^2 \times S^2$ and $S^2 \widetilde{\times} S^2$ (the twisted S^2 bundle over S^2). Similarly, we can obtain $f^{-1}(3/2)$ from M_{L_2} by adding copies of $S^2 \times S^2$ and $S^2 \widetilde{\times} S^2$. Since

$$(S^2 \times S^2) \# \mathbb{CP}^2 \approx 2\mathbb{CP}^2 \# \overline{\mathbb{CP}}^2 \text{ and } S^2 \widetilde{\times} S^2 \approx \mathbb{CP}^2 \# \overline{\mathbb{CP}}^2,$$

we can achieve connected summing with $S^2 \times S^2$ or $S^2 \widetilde{\times} S^2$ using moves (i) and (ii); see Exercise 6.3. Hence, we have shown that we can change L_1 and L_2 using Kirby moves (i) and (ii) such that $M_{L_1} = M_{L_2}$. Let us write $M := M_{L_1} = M_{L_2}$.

Step 2. The framed links L_1 and L_2 give two handlebody structures on M, which correspond to Morse functions $g_i \colon M \to [-1,1]$ for $i \in \{1,2\}$ such that $g_i^{-1}(-1)$ is the only critical point of index zero, $g_i^{-1}(0) = S^3$, and $g_i^{-1}(1) = \partial M$. Furthermore, all critical points of index two have values in $(0,1)$. After possibly taking connected sums of M with some number of copies of $S^2 \times S^2$, using Cerf theory, we are going to construct a one-parameter family of Morse functions $\{g_t : t \in [1,2]\}$ connecting g_1 and g_2, where the only possible bifurcation is two index two critical points having the same value. The intersections of the stable manifolds of the index two critical points with $S^3 = g_t^{-1}(0)$ change by an isotopy, except when there is a flow line between two index two critical points. This corresponds to a handle slide.

The Cerf-theoretical argument is analogous to the proof of the Reidemeister–Singer theorem for Heegaard diagrams (Theorem 3.35). It follows from Lemma 2.76 that there is a generic one-parameter family g_t which is an ordered Morse function for generic $t \in [1,2]$, and there is a single index zero critical point and no index four critical points for every $t \in [1,2]$.

Next, we remove all critical points of index one and three from the family g_t. Using the Beak move and the Independent Trajectories Principle, we move all index one-two births to the beginning and index one-two deaths to the end. We now eliminate all crossings between index one critical points. Consider the first such crossing c. If we have a flow line between the two index one critical points before c, then we can remove it in a neighbourhood of a beak using the following lemma:

Lemma 6.5 *If an i-handle and an $(i+1)$-handle are born and then the i-handle slides over another i-handle, then we can instead delay the birth until after the handle slide; see Figure 6.4.*

Proof This corresponds to a codimension-two bifurcation where there is a gradient flow line from an index i critical point to an index i-$(i+1)$ birth singularity. The bifurcation diagram in $\mathbb{R}^2$ consists of a birth stratum that is the x-axis, and a handle slide stratum that is the positive y-axis. If a one-parameter

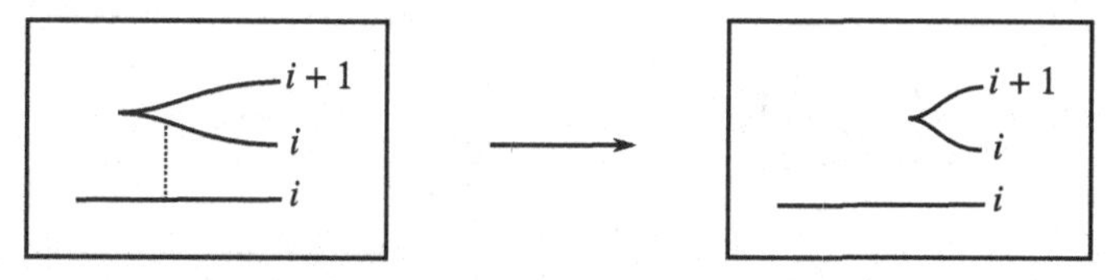

Figure 6.4 Removing a flow line between index i critical points following an index i-$(i+1)$ birth.

family crosses the birth stratum along the negative x-axis followed by crossing the handle slide stratum, we can instead go around the the origin the other way, only crossing the birth stratum along the positive x-axis. □

We then apply the Beak move and the Independent Trajectories Principle repeatedly to exchange c with a crossing between index two critical points; see the top row of Figure 6.5. This way, we can arrange that there is no crossing between index one critical points; see the bottom left of Figure 6.5. We then move the first index one-two birth and the last index one-two death off the edge of the graphic, as in the bottom right of Figure 6.5.

We connect in S^3 the endpoints of the core of the handle corresponding to an index one critical point, obtaining a closed curve S_t^1 in M. Since $\pi_1(M) = 1$, the normal bundle of S_t^1 in M has a unique trivialisation. If we perform surgery along S_t^1, we obtain $M \# (S^2 \times S^2)$. In fact, Kirby shows that the surgery can be performed continuously. This way, one can surger out all index one critical points. If we replace g_t by $-g_t$, we can also remove the index three critical points. This leaves us with only index two critical points. This shows that, after possibly taking connected sums with $S^2 \times S^2$, which we can arrange with Kirby moves (i) and (ii), one can get from L_1 to L_2 using only handle slides. □

There is also a Kirby calculus for four-manifolds. Let X be a smooth, connected four-manifold, and f_0, f_1 two ordered Morse functions on it with a single index zero critical point. Then Lemma 2.76 implies that we can connect them with a generic one-parameter family of smooth functions f_t for $t \in I$ such that every f_t is ordered for generic t and has a unique index zero critical point. If we consider the Kirby diagrams of X corresponding to f_t, they change using the following moves:

(i) An isotopy of the one-handles and two-handles.
(ii) A one-handle slides over a one-handle, or a two-handle slides over a one-handle or two-handle.
(iii) An index one-two or two-three birth or death.

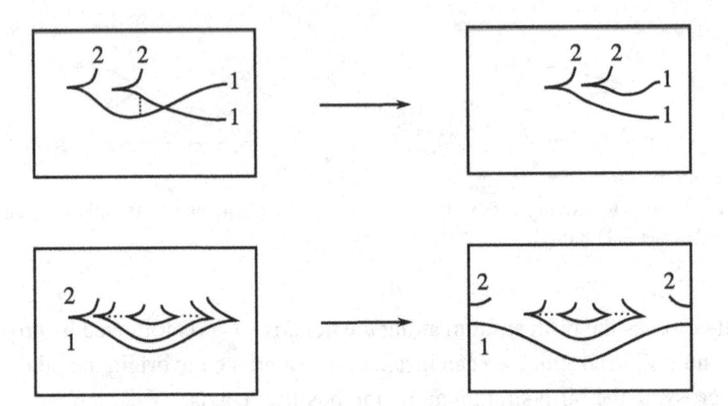

Figure 6.5 The top row shows how to exchange two index one-two births in order to eliminate the first intersection point between index one critical points. This way, we can obtain the configuration in the bottom left. By moving the first index one-two birth and the last index one-two death off the edge of the graphic, we obtain the graphic in the bottom right.

In dotted circle notation, an index one-two birth results in an unknotted two-handle, which is linked by a dotted circle. In general, one can cancel a one-handle and a two-handle whenever the two-handle passes through the one-handle exactly once.

When we are considering Kirby diagrams of closed four-manifolds; that is, the boundary of the two-handlebody is a connected sum of copies of $S^1 \times S^2$ that we fill with three-handles and a four-handle, an index two-three birth corresponds to the appearance of an unknotted, zero-framed two-handle.

For an in-depth study of four-manifold topology from the point of view of Kirby calculus, see the books of Kirby [85], Gompf and Stipsicz [48], and Akbulut [6].

6.2 The Intersection Form

Let X be a closed, connected, oriented, and simply connected four-manifold. Then $H_1(X) \cong H_3(X) \cong 0$. Hence $H_2(X)$ and $H^2(X)$ are free Abelian groups by the Universal Coefficient Theorem. We denote the fundamental class of X by $[X] \in H_4(X)$. Then the *intersection form*

$$Q_X \colon H^2(X) \times H^2(X) \to \mathbb{Z}$$

of X is given by

$$Q_X(a, b) := \langle a \cup b, [X] \rangle$$

for a, $b \in H^2(X)$. This is a non-degenerate, symmetric bilinear form. It is *unimodular*; that is, it induces an isomorphism

$$H^2(X) \to \mathrm{Hom}(H^2(X), \mathbb{Z}) \cong H_2(X)$$

by Poincaré duality. Hence, its matrix in any basis is invertible over $\mathbb{Z}$; that is, it has determinant ± 1.

Proposition 6.6 *Let X be a closed, oriented four-manifold, and $s \in H^2(X)$ a cohomology class. Then there exists an oriented surface S embedded in X such that $[S] \in H_2(X)$ is Poincaré dual to s.*

Proof Since the Eilenberg–MacLane space $K(\mathbb{Z}, 2)$ is $\mathbb{CP}^\infty$, we have

$$H^2(X) \cong [X, \mathbb{CP}^\infty].$$

Then s corresponds to a continuous map $f_s \colon X \to \mathbb{CP}^\infty$ such that $f_s^* x = s$, where x is the generator of $H^2(\mathbb{CP}^\infty) \cong \mathbb{Z}[x]$, which is represented by the hypersurface $\mathbb{CP}^{\infty-1}$. We can homotope f such that it becomes smooth and transverse to $\mathbb{CP}^{\infty-1}$. Then we let $S := f_s^{-1}(\mathbb{CP}^{\infty-1})$. □

If A and B are surfaces Poincaré dual to a, $b \in H^2(X)$ (these exist by Proposition 6.6), then $Q_X(a, b)$ is the algebraic intersection number $[A] \cdot [B]$ of A and B. This justifies the term 'intersection form'.

We can easily read off the intersection form of a four-manifold from its Kirby diagram. Suppose that the four-manifold X is given by a Kirby diagram without one-handles; that is, a framed link $L = L_1 \cup \cdots \cup L_n$ in S^3 with framing coefficient $r_i \in \mathbb{Z}$ for $i \in \{1, \dots, n\}$. Then we obtain a basis $B_1, \dots, B_n$ of $H_2(X)$ by capping off a Seifert surface of L_i with the core of the two-handle attached along L_i. Furthermore, $B_i \cdot B_j = \mathrm{lk}(L_i, L_j)$ if $i \neq j$ and $B_i \cdot B_i = r_i$ by Lemma 4.27.

The following result is due to Milnor and Whitehead:

Theorem 6.7 *Let X and X' be closed, connected, and simply connected four-manifolds. Then X and X' are homotopy equivalent if and only if $Q_X \cong Q_{X'}$.*

Proof We follow the exposition of Kirby [85]. Let $X_0 := X \setminus B^4$. Then $H_0(X_0) \cong \mathbb{Z}$, $H_1(X_0) \cong 0$, $H_2(X_0) \cong \mathbb{Z}^n$ for some $n \in \mathbb{N}$ (it is torsion-free by the Universal Coefficient Theorem), and $H_i(X_0) = 0$ for $i > 2$. Since $\pi_1(X_0) \cong 1$, the Hurewicz map $\pi_2(X_0) \to H_2(X_0)$ is an isomorphism. Let $f_i \colon S^2 \to X_0$ be based maps for $i \in \{1, \dots, n\}$ such that their homotopy classes $[f_1], \dots, [f_n]$ form a basis of $\pi_2(X_0)$. Let

$$f := f_1 \vee \cdots \vee f_n \colon S^2 \vee \cdots \vee S^2 \to X_0.$$

Then f induces an isomorphism on homology, and since both spaces are simply connected, it is a homotopy equivalence by Whitehead's theorem. So,

$$X \cong (S^2 \vee \cdots \vee S^2) \cup_g D^4$$

for a gluing map $g\colon \partial D^4 \to S^2 \vee \cdots \vee S^2$.

Hence, the homotopy type of X only depends on the gluing map g. We now show that this is uniquely determined by Q_X. We can homotope g such that it is smooth in the complement of the pre-image of the wedge point. By the Pontryagin construction (Section 2.7), the homotopy class of g is determined by the framed cobordism class of the link $L = L_1 \cup \cdots \cup L_n$, where $L_i := g^{-1}(p_i)$ for a regular value p_i of g in the ith S^2 component of $S^2 \vee \cdots \vee S^2$, and the framing is given by pulling back a tangent frame in $T_{p_i}S^2$ for $i \in \{1, \ldots, n\}$. Framed cobordism has to preserve the labelling of the components L_i in the sense that a cobordism C from $L_1 \cup \cdots \cup L_n$ to $L_1' \cup \cdots \cup L_n'$ can be written as $C = C_1 \cup \cdots \cup C_n$, where C_i is a framed cobordism from L_i to L_i'.

We can make each L_i connected using a framed cobordism that corresponds to taking the connect sum of the components of L_i. As in the paragraph preceding the statement of Theorem 6.7, the linking matrix of L is isomorphic to the intersection form of X. Indeed, if we choose a Seifert surface for L_i, the gluing map g collapses its boundary to the point p_i, giving rise to a two-cycle B_i in X dual to $(f_i)_*([S^2])$. Then $[B_i] \cdot [B_j] = \mathrm{lk}(L_i, L_j)$ for $i \neq j$ by Lemma 4.27, and $[B_i] \cdot [B_i]$ is the framing coefficient of L_i.

What remains to be shown is that the linking matrix uniquely determines the labelled, framed cobordism class of L. It follows from Lemma 4.27 that labelled, framed cobordant links have the same linking matrix. Conversely, when $n = 1$, the framed cobordism class of L is determined by the framing coefficient according to the computation of $\pi_2(S^3)$ in Remark 2.30. We now proceed using induction on n. Suppose that $L = L_1 \cup \cdots \cup L_n$ and $L' = L_1' \cup \cdots \cup L_n'$ are framed links with the same linking matrix $A = (a_{ij})$. By the inductive hypothesis, we can arrange that $L_i = L_i'$ for $i \in \{1, \ldots, n-1\}$. We can apply band moves to both L_n and L_n' such that they become the union of a_{in} meridians of L_i for $i \in \{1, \ldots, n-1\}$, and all have framing zero except one meridian of L_1 with framing a_{nn}. It follows that L and L' are labelled, framed cobordant. □

We can also consider the intersection form Q_X on $H^2(X; \mathbb{R})$. Non-degenerate symmetric bilinear forms over *real* vector spaces are completely determined by their rank and signature. We write $b_2^+(X)$ and $b_2^-(X)$ for the dimensions of maximal positive and negative definite subspaces of $H^2(X; \mathbb{R})$, respectively. The signature of Q_X is

$$\sigma(X) = b_2^+(X) - b_2^-(X).$$

Signature is a cobordism invariant, and it vanishes if and only if X is oriented null-cobordant; see Section 1.7.

The intersection form Q_X is positive definite if $b_2^+(X) = b_2(X)$, negative definite if $b_2^-(X) = b_2(X)$, and is called *indefinite* otherwise.

Definition 6.8 The intersection form Q_X is *even* if $Q(x, x)$ is even for every $x \in H^2(X)$, and is *odd* otherwise. We call this the *type* of X.

Definition 6.9 The Lie group $\mathrm{Spin}(n)$ is the connected double cover of $\mathrm{SO}(n)$ for $n \geq 2$. The oriented n-manifold M is *spin* if the structure group of TM can be lifted from $\mathrm{SO}(n)$ to $\mathrm{Spin}(n)$. A *spin structure* on M is an equivalence class of such lifts.

Proposition 6.10 *The intersection form Q_X is even if and only if X is spin.*

Proof This is because the Spin structure of D^4 extends to a two-handle if and only if the framing of the attaching sphere is even, and the self-intersection of the corresponding element of $H_2(X)$ is precisely the framing. □

Proposition 6.11 (Adjunction formula) *Let C be a nonsingular, connected complex curve in the complex four-manifold X. Then*

$$2g(C) - 2 = [C] \cdot [C] - \langle c_1(X), [C] \rangle.$$

Proof If ν_C denotes the normal bundle of C in X, then $TX|_C \cong TC \oplus \nu_C$. Hence, using the Whitney sum formula,

$$c_1(TX|_C) = c_1(TC) + c_1(\nu_C) = e(TC) + e(\nu_C).$$

By the naturality of Chern classes, $\langle c_1(X), [C] \rangle = \langle c_1(TX|_C), [C] \rangle$. So

$$\langle c_1(X), [C] \rangle = \langle e(TC), [C] \rangle + \langle e(\nu_C), [C] \rangle = \chi(C) + [C] \cdot [C],$$

where the second equality follows from the obstruction-theoretic interpretation of the Euler class as the dual of the intersection of the zero-section of the bundle with a generic section. Since $\chi(C) = 2 - 2g(C)$, the result follows. □

Corollary 6.12 *If C is a smooth algebraic curve in $\mathbb{CP}^2$ of degree d, then*

$$g(C) = \frac{(d-1)(d-2)}{2}.$$

Proof The curve C intersects a generic line $\mathbb{CP}^1$ algebraically in d points by the fundamental theorem of algebra. Hence $[C] = d[\mathbb{CP}^1] \in H_2(\mathbb{CP}^2)$. Note that $c_1(\mathbb{CP}^2) = 3\text{PD}([\mathbb{CP}^1])$. The adjunction formula now implies that

$$2g(C) - 2 = d^2 - 3d,$$

and the result follows. □

Exercise 6.13 Let X be an oriented four-manifold and $A \in H_2(X)$. Show that

$$\langle w_2(X), A \rangle \equiv A \cdot A \mod 2.$$

(Hint: Use the idea of the proof of Proposition 6.11.) Combined with Proposition 6.10, show that this implies that X is spin if and only if $w_2(X) = 0$.

Indefinite intersection forms have a simple classification:

Theorem 6.14 *Two indefinite, unimodular, symmetric bilinear forms over $\mathbb{Z}$ are equivalent if and only if they have the same rank, signature, and type.*

Not every value of the triple rank, signature, and type can be realised:

Proposition 6.15 *The signature of an even form is divisible by eight.*

However, this is the only restriction for indefinite forms. An important example of a definite even form of signature eight is the E_8 form, given by the matrix

$$E_8 = \begin{pmatrix} 2 & 1 & 0 & 0 & 0 & 0 & 0 & 0 \\ 1 & 2 & 1 & 0 & 0 & 0 & 0 & 0 \\ 0 & 1 & 2 & 1 & 0 & 0 & 0 & 0 \\ 0 & 0 & 1 & 2 & 1 & 0 & 0 & 0 \\ 0 & 0 & 0 & 1 & 2 & 1 & 0 & 1 \\ 0 & 0 & 0 & 0 & 1 & 2 & 1 & 0 \\ 0 & 0 & 0 & 0 & 0 & 1 & 2 & 0 \\ 0 & 0 & 0 & 0 & 1 & 0 & 0 & 2 \end{pmatrix}. \tag{6.1}$$

The four-manifold P_{E_8} with boundary defined by the Kirby diagram in Figure 6.6 has intersection form E_8. Its boundary is the Poincaré homology sphere Σ_P, which is the Brieskorn manifold $M(2,3,5)$; see Definition 2.66. By the work of Freedman [37], there is a topological four-manifold Δ with boundary Σ_P that has the same homotopy groups as D^4. The E_8 *manifold* $M_{E_8} := P_{E_8} \cup \Delta$ is a closed topological four-manifold with intersection form E_8.

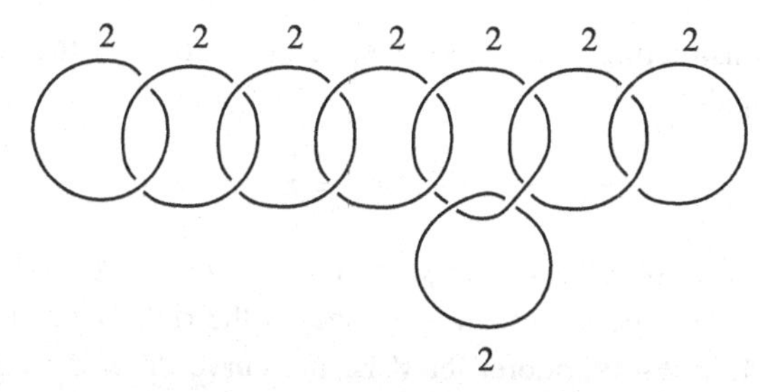

Figure 6.6 A Kirby diagram for a four-manifold whose intersection form is the E_8 lattice.

The hyperbolic form

$$H = \begin{pmatrix} 0 & 1 \\ 1 & 0 \end{pmatrix}$$

has rank two and signature zero. (Note that $Q_{S^2 \times S^2} = H$.) Hence, every symmetric, unimodular, indefinite, even form is isomorphic to $kE_8 \oplus lH$ for some $k \in \mathbb{Z}$ and $l \in \mathbb{Z}_+$. On the other hand, every indefinite odd form is isomorphic to $k(1) \oplus l(-1)$ for $k, l \in \mathbb{Z}_+$.

Not every symmetric, unimodular bilinear form can be the intersection form of a *smooth* four-manifold, due to the following result of Rokhlin [154]:

Theorem 6.16 *Let X be a closed, smooth, spin four-manifold (i.e., Q_X is even). Then $\sigma(X)$ is divisible by 16.*

If M is a closed topological n-manifold, then the *Kirby–Siebenmann class* $\kappa(M) \in H^4(M; \mathbb{Z}_2)$ vanishes if M admits a PL structure. When $\dim(M) \geq 5$, the converse also holds. For further details, see Kirby and Siebenmann [87]. When X is a spin four-manifold,

$$\kappa(X) \equiv \sigma(X)/8 \mod 2.$$

A topological four-manifold has a PL structure if and only if it has a smooth structure, and so a smooth four-manifold has $\kappa(X) = 0$.

Example 6.17 An example of a smooth spin four-manifold with signature -16 is the *K3 surface*. It can be constructed as the solution to

$$x^4 + y^4 + z^4 + w^4 = 0$$

in $\mathbb{CP}^3$. Its intersection form is $-2E_8 \oplus 3H$.

From the point of view of complex geometry and algebraic geometry, there are many different K3 surfaces, but they are all diffeomorphic by the work of Kodaira. We now give a description of a K3 surface that will be particularly useful when constructing exotic smooth structures. Consider the homogeneous cubic polynomials

$$p_0 := x^3 - y^2 z \text{ and } p_1 := x^3 + y^3 + z^3.$$

The equation $p_0 = 0$ defines a singular curve $C_2 \subset \mathbb{CP}^2$ called a *cusp*. This has a single singular point, which is a cone on the right-handed trefoil knot; see Exercise 4.42. Furthermore, let V be the curve defined by $p_1 = 0$ and $R := C_2 \cap V$. By Bézout's theorem, $|R| = 9$. For $[t_0 : t_1] \in \mathbb{CP}^1$, let

$$p_{[t_0:t_1]} := t_0 p_0 + t_1 p_1.$$

Then $\{p_b : b \in \mathbb{CP}^1\}$ is a *pencil of curves*. Any two distinct curves in the pencil intersect in R. For $a \in \mathbb{CP}^2 \setminus R$, we let $\pi(a)$ be the unique point b of $\mathbb{CP}^1$ for which a lies on p_b. For generic $[t_0 : t_1] \in \mathbb{CP}^1$, the curve $p_b = 0$ is a smooth cubic curve, so, by Corollary 6.12, topologically a torus. If we blow up $\mathbb{CP}^1$ at the points of R, we can extend π to a fibration

$$\pi \colon \mathbb{CP}^2 \# 9\overline{\mathbb{CP}}^2 \to \mathbb{CP}^1$$

whose generic fibres are smooth elliptic curves. This is an instance of an *elliptic fibration*, and we denote the resulting complex curve by $E(1)$.

We now take the *fibre sum* of two copies of $E(1)$: For a generic $b \in \mathbb{CP}^1$, let $N(F) \approx D^2 \times T^2$ be a regular neighbourhood of the fibre $F := \pi^{-1}(b)$, and we glue together two copies of $E(1) \setminus N(F)$ using a fibre-preserving and orientation-reversing diffeomorphism ϕ between the boundary three-tori. Since π has a cusp fibre, the diffeomorphism type of the resulting four-manifold $E(2)$ is independent of ϕ. By construction, $E(2)$ admits an elliptic fibration over $\mathbb{CP}^1$. Then $E(2)$ is a K3 surface.

It follows from Rokhlin's theorem (Theorem 6.16) that a closed topological four-manifold with intersection form E_8 (e.g., M_{E_8}) does not admit a smooth structure. By a celebrated result of Freedman [37][38], the pair $(Q_X, \kappa(X))$ is a complete invariant of topological, simply connected four-manifolds X:

Theorem 6.18 *Two closed, simply connected four-manifolds are homeomorphic if and only if they have isomorphic intersection forms and the same Kirby–Siebenmann class.*

Conversely, for any symmetric, unimodular bilinear form Q over $\mathbb{Z}$, there is a unique topological four-manifold X with $Q_X = Q$ if Q is even, and two if Q is odd, distinguished by $\kappa(X)$.

Note that, when Q_X is odd and $\kappa(X) = 1$, then X does not admit a smooth structure. However, $\kappa(X) = 0$ does not guarantee that X is smoothable.

The key to the proof of Theorem 6.18 is the Whitney trick in the topological category, which in turn implies the h-cobordism theorem just like in dimensions five and higher. What Freedman showed is that, in a simply connected four-manifold, a Whitney disk can be realised as a topologically locally flat submanifold. In contrast, in the smooth category, it is an immersion with transverse double points. For a recent exposition, see Behrens et al. [9].

In Theorem 6.14, we saw that the indefinite forms have a simple classification. On the other hand, there is a wild world of definite forms. The following surprising result of Donaldson [29] implies that most of these forms do not arise as intersection forms of *smooth* four-manifolds:

Theorem 6.19 *Let X be a closed, smooth, simply connected four-manifold such that Q_X is positive (or negative) definite. Then Q_X is diagonalisable; that is, isomorphic to $n(1)$ (or $n(-1)$).*

Combined with Freedman's theorem, we obtain that every smooth, definite four-manifold is homeomorphic to either $\#_n\mathbb{CP}^2$ or $\#_n\overline{\mathbb{CP}}^2$.

By Theorem 6.14, indefinite forms are determined by their rank, signature, and type. In particular, indefinite odd forms are of the form $k(1) \oplus l(-1)$, so every four-manifold with such an intersection form is homeomorphic to $k\mathbb{CP}^2 \# l\overline{\mathbb{CP}}^2$ for $k, l > 0$.

What remains open is the geography problem for smooth, indefinite, even (i.e., spin) four-manifolds. Namely, which of these intersection forms are represented by smooth four-manifolds. Up to homeomorphism, they are all determined by their rank and signature. And we have seen that they have intersection forms $2kE_8 \oplus lH$ for $k \in \mathbb{Z}$ and $l > 0$. We can suppose that $k \geq 0$ by possibly reversing the orientation of the manifold. If $l \geq 3k$, then $2kE_8 \oplus lH$ is the intersection form of $kK3 \# (l - 3k)(S^2 \times S^2)$. The condition $l \geq 3k$ is equivalent to $b_2(X) \geq \frac{11}{8}\sigma(X)$. The $\frac{11}{8}$-conjecture, due to Furuta, states that this condition is also necessary for $2kE_8 \oplus lH$ to be the intersection form of a smooth, simply connected four-manifold. The best known result, due to Furuta [43], states that $b_2(X) \geq \frac{10}{8}\sigma(X)$ for every smooth, indefinite, even four-manifold.

Another difficult question is the classification of smooth structures up to diffeomorphism on a given topological four-manifold. We have seen that in higher dimensions, there are manifolds that admit more than one smooth structure, but the number is always finite if the manifold is closed. In dimensions below four, every manifold admits a unique smooth structure. In contrast, Taubes has shown that $\mathbb{R}^4$ admits continuum many non-diffeomorphic smooth structures. Note that $\mathbb{R}^n$ has a unique smooth structure for $n \neq 4$.

If X is a closed topological four-manifold, then it can be represented by a finite Kirby diagram, and hence admits at most countably infinite pairwise non-diffeomorphic smooth structures. The K3 surface, for example, admits infinitely many smooth structures, and so does $\mathbb{CP}^2 \# k\overline{\mathbb{CP}}^2$ for $k \geq 3$. There is no known example of a smooth four-manifold that admits only finitely many smooth structures. One of the most difficult open problems in topology is the smooth four-dimensional Poincaré conjecture, which asks whether S^4 admits a unique smooth structure.

For a thorough and entertaining overview of four-manifold topology, see the book of Scorpan [159].

6.3 The Seiberg–Witten Invariant

To distinguish smooth structures, the main tools are the Seiberg–Witten invariant and the conjecturally equivalent Ozsváth–Szabó four-manifold invariant defined using Heegaard Floer homology.

We first need to discuss Spin^c structures on four-manifolds. Recall that we introduced Spin^c structures on three-manifolds in Definition 5.4.

Exercise 6.20 Let $\mathrm{Sp}(1) \approx S^3$ be the group of unit quaternions, and consider the action of $\mathrm{Sp}(1) \times \mathrm{Sp}(1)$ on the quaternions $\mathbb{H}$ given by $(q, q') \cdot r := qrq'^{-1}$ for $q, q' \in \mathrm{Sp}(1)$ and $r \in \mathbb{H}$. Show that this action is via special orthogonal transformations of $\mathbb{H}$, and the induced map $\mathrm{Sp}(1) \times \mathrm{Sp}(1) \to \mathrm{SO}(4)$ is a double cover. By identifying $\mathrm{Sp}(1)$ with $\mathrm{SU}(2)$, conclude that

$$\mathrm{Spin}(4) \cong \mathrm{SU}(2) \times \mathrm{SU}(2).$$

In addition to the failure of the Whitney trick, the splitting of Spin(4) in Exercise 6.20 is another key reason for the exotic phenomena in dimension four.

Consider the *complex spin group*

$$\mathrm{Spin}^c(4) := \mathrm{Spin}(4) \times_{\mathbb{Z}_2} U(1),$$

where $\mathbb{Z}_2$ acts on the two factors diagonally by multiplication with ± 1. Projection from $\mathrm{Spin}(4) \times U(1)$ to the first factor composed with the covering map $\mathrm{Spin}(4) \to \mathrm{SO}(4)$ descends to a group homomorphism

$$\phi \colon \mathrm{Spin}^c(4) \to \mathrm{SO}(4).$$

Analogously to spin structures (Definition A.14), we can define Spin^c structures as follows:

Definition 6.21 Let X be an oriented Riemannian four-manifold. We write $p\colon P(TX) \to X$ for the associated principal SO(4)-bundle consisting of positive orthonormal four-frames in TX. A *lift of the structure group of X to $Spin^c(4)$* is a pair

$$(\pi\colon P \to X, \epsilon\colon P \to P(TX)),$$

where π is a principal $\mathrm{Spin}^c(4)$-bundle and ϵ is a bundle morphism that is fibrewise the covering ϕ defined earlier, and such that $p \circ \epsilon = \pi$. We say that the lifts $(\pi\colon P \to X, \epsilon)$ and $(\pi'\colon P' \to X, \epsilon')$ are *equivalent* if there is a bundle isomorphism $\iota\colon P \to P'$ such that $\epsilon' \circ \iota = \epsilon$; that is, the following diagram is commutative:

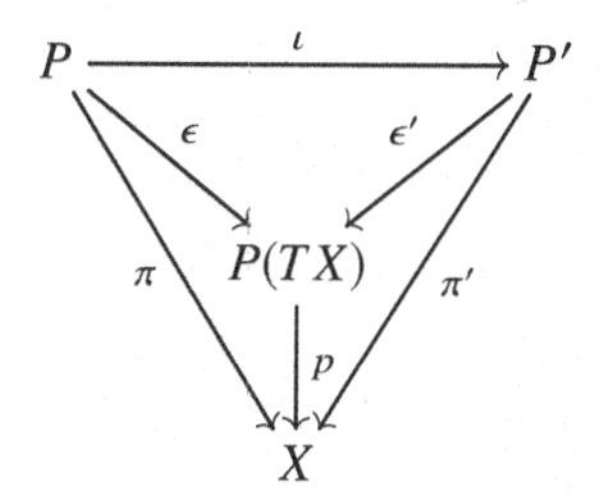

A *Spin^c structure* on X is an equivalence class of lifts of the structure group SO(4) to $\mathrm{Spin}^c(4)$. We denote by $\mathrm{Spin}^c(X)$ the set of Spin^c structures on X.

A Spin^c structure on X is equivalent to an almost complex structure over the two-skeleton of X that extends to the three-skeleton. Phrased more in line with Definition 5.4, it is the homology class of an almost complex structure J on $X \setminus H$ for a finite subset $H \subset \mathrm{Int}(X)$, where (J, H) and (J', H') are *homologous* if there is a compact one-manifold $C \subset X$ containing H and H' such that $J|_{X\setminus H}$ and $J'|_{X\setminus H}$ are isotopic through almost complex structures.

Suppose that X has boundary and $\mathfrak{s} \in \mathrm{Spin}^c(X)$ is represented by an almost complex structure J on $X \setminus H$ for $H \subset \mathrm{Int}(X)$ finite. Let

$$\xi := T(\partial X) \cap JT(\partial X)$$

be the bundle of J-complex lines tangent to ∂X. Then the homology class of the normal vector field $\xi^\perp$ gives the restriction

$$\mathfrak{s}|_{\partial X} \in \mathrm{Spin}^c(\partial X)$$

in the sense of Definition 5.4.

By obstruction theory, X admits a Spin^c structure if and only if $w_2(X) \in H^2(X;\mathbb{Z}_2)$ lifts to a class in $H^2(X;\mathbb{Z})$, which is always the case when X is compact by work of Hirzebruch and Hopf. Furthermore, when $H_1(X)$ has no two-torsion, Spin^c structures correspond to integral lifts of $w_2(X)$. One can associate a complex line bundle $\det(\mathfrak{s})$ to a Spin^c structure $\mathfrak{s}$ via the representation $\mathrm{Spin}^c(4) \to U(1)$, called the *determinant line bundle of* $\mathfrak{s}$. Then

$$c_1(\det(\mathfrak{s})) \equiv w_2(X) \mod 2$$

is the lift of $w_2(X)$ corresponding to the Spin^c structure $\mathfrak{s}$.

Let X be a smooth, closed, connected, and oriented four-manifold. We also fix an orientation of

$$H^0(X;\mathbb{R}) \otimes \det H^1(X;\mathbb{R}) \otimes \det H^2_+(X;\mathbb{R}),$$

called a *homology orientation* of X. The Seiberg–Witten invariant is defined when $b_2^+(X) > 0$, and is a function

$$SW_X\colon \mathrm{Spin}^c(X) \to \mathbb{Z}.$$

If we do not fix a homology orientation, the map is only well defined up to sign. When $b_2^+(X) = 1$, the invariant also depends on the choice of one of two chambers of the space of Riemannian metrics on X; hence, we will assume that $b_2^+(X) > 1$ from now on.

For the definition of SW_X, see Morgan [121]. Here, we treat it as a black box and only summarise its key properties. Firstly, the support

$$\mathcal{B}_X \subset \mathrm{Spin}^c(X) \cong H^2(X)$$

of SW_X is finite and is called the set of *basic classes*. If $\beta \in \mathcal{B}_X$, then

$$SW_X(-\beta) = (-1)^{(\chi(X)+\sigma(X))/4} SW_X(\beta),$$

so $-\beta \in \mathcal{B}_X$. Hence, we can write

$$\mathcal{B}_X \setminus \{0\} = \{\pm\beta_1, \dots, \pm\beta_n\}.$$

We can view SW_X as an element $SW(X)$ of the group ring $\mathbb{Z}[H^2(X)]$ as follows. Let $t_\beta := \exp(\beta)$ for $\beta \in H^2(X)$, with the relations $t_{\alpha+\beta} = t_\alpha t_\beta$. Then

$$SW(X) := SW_X(0) + \sum_{j=1}^{n} SW_X(\beta_j)\left(t_{\beta_j} + (-1)^{(\chi(X)+\sigma(X))/4} t_{\beta_j}^{-1}\right).$$

If X has a metric of positive scalar curvature, or if $X = X_1 \# X_2$, where $b_2^+(X_1) > 0$ and $b_2^+(X_2) > 0$, then $SW(X) = 0$. On the other hand, we have the following nonvanishing result of Taubes [166]:

Theorem 6.22 *Let X be a closed, connected four-manifold with $b_2^+(X) > 1$. Let ω be a symplectic form on X and orient X with $\omega \wedge \omega$. Then the Spin^c structure of the almost complex structure compatible with ω has Seiberg–Witten invariant ± 1.*

If X is simply connected, then we say that it is of *simple type* if

$$\beta^2 = c_1^2(X) = 3\sigma(X) + 2\chi(X)$$

for every basic class $\beta \in \mathcal{B}_X$. If X has a symplectic structure, it is of simple type. There is no known example of a simply connected four-manifold X with $b_2^+(X) > 1$ that is not of simple type.

The following describes how the set of basic classes changes under blow-ups:

Theorem 6.23 *Let X be a simply connected four-manifold of basic type. If $X' = X\#\overline{\mathbb{CP}}^2$ is the blow-up of X and $E \in H^2(X')$ is the dual of the exceptional divisor, then*

$$\mathcal{B}_{X'} = \{\beta \pm E : \beta \in \mathcal{B}_X\}.$$

We have the following generalisation of Proposition 6.11 due to Kronheimer and Mrowka [90]:

Theorem 6.24 (Adjunction inequality) *Let X be a simply connected four-manifold. If S is an embedded, closed, connected, oriented surface in X with nonnegative self-intersection, then*

$$2g(S) - 2 \geq [S] \cdot [S] + |\beta([S])|$$

for every $\beta \in \mathcal{B}_X$.

Kronheimer and Mrowka used this to prove the Thom conjecture:

Theorem 6.25 *Let S be an embedded, closed, connected, and oriented surface in $\mathbb{CP}^2$ such that $[S] = d[\mathbb{CP}^1] \in H_2(\mathbb{CP}^2)$. Then*

$$g(S) \geq \frac{(d-1)(d-2)}{2}.$$

Proof Since $\mathbb{CP}^2$ is Kähler and hence symplectic, $SW(c_1(\mathbb{CP}^2)) = \pm 1$ by Theorem 6.22. The result now follows from Theorem 6.24 and the fact that $c_1(\mathbb{CP}^2) = 3\mathrm{PD}([\mathbb{CP}^1])$. □

By Corollary 6.12, this means that, among smoothly embedded surfaces in a given homology class in $\mathbb{CP}^2$, complex curves have minimal genus.

Suppose that $b_2^+(X) > 1$ and $T \subset X$ is a smoothly embedded, homologically essential torus with trivial self-intersection, and let $K \subset S^3$ be a knot. Fintushel and Stern [33] defined the *knot surgery* operation on X, resulting in the four-manifold X_K. This is obtained by gluing $X \setminus N(T)$ and $S^1 \times (S^3 \setminus N(K))$ via an orientation-reversing diffeomorphism of their boundaries that maps a meridian of T to a longitude of K. There is a canonical identification between $H^2(X)$ and $H^2(X_K)$, and hence between the group rings $\mathbb{Z}[H^2(X)]$ and $\mathbb{Z}[H^2(X_K)]$. Let $t := \exp(2\mathrm{PD}[T])$, where $[T]$ is the homology class induced by T in $H_2(X_K)$. Fintushel and Stern showed the following:

Theorem 6.26 *Let X be a simply connected, smooth, closed, connected, and oriented four-manifold such that $b_2^+(X) > 1$. If X_K is obtained from X via knot surgery along the torus T, as earlier, then*

$$SW(X_K) = \Delta_K(t) \cdot SW(X), \tag{6.2}$$

where $\Delta_K(t)$ is the symmetrised Alexander polynomial of K.

If $\pi_1(X \setminus T) \cong 1$, then X and X_K are simply connected and have the same intersection form, and are hence homeomorphic by Freedman's theorem (Theorem 6.18). We say that p is an A-polynomial if it is a symmetric integral Laurent polynomial satisfying $p(1) = \pm 1$. Every A-polynomial is the symmetrised Alexander polynomial of a knot in S^3. Consequently, if $SW(X) \neq 0$, then we obtain infinitely many pairwise non-diffeomorphic smooth structures on X corresponding to A-polynomials.

When X is the K3 surface $E(2)$ defined in Example 6.17, it is Kähler and hence symplectic. So, by Theorem 6.22, we have $SW(X) \neq 0$ (actually, $SW(X) = 1$). If T is a regular torus fibre of the elliptic fibration $\pi\colon E(2) \to \mathbb{CP}^1$ close to a cusp fibre, then $\pi_1(X \setminus T) = 1$. Hence, we obtain a different smooth structure on X for every A-polynomial $p(t)$ by performing knot surgery on $E(2)$ along T using a knot K with $\Delta_K(t) = p(t)$.

Morgan [121] covers the basics of Seiberg–Witten theory, while Gompf and Stipsicz [48] and Scorpan [159] review numerous applications.

Appendix
Fibre Bundles and Characteristic Classes

Here, we review the necessary background on fibre bundles and characteristic classes without proofs. For further details, see Steenrod [165] and Milnor–Stasheff [118].

A.1 Fibre Bundles

Intuitively, a fibre bundle is a space E that is locally a product but might be twisted globally.

Definition A.1 A *fibre bundle* consists of a surjective, continuous map $\pi\colon E \to B$, where E is called the *total space* and B the *base space*, and a topological space F called the *fibre*. We assume that B is path-connected, and that each point of B has a neighbourhood U such that there is a homeomorphism $h\colon \pi^{-1}(U) \to U \times F$ that makes the following diagram commutative:

$$\begin{array}{ccc} \pi^{-1}(U) & \xrightarrow{h} & U \times F, \\ {\scriptstyle\pi}\downarrow & \swarrow {\scriptstyle p_U} & \\ U & & \end{array}$$

where p_U is projection onto U.

It immediately follows from the definition that $F_b := \pi^{-1}(\{b\})$ is homeomorphic to the fibre F for every $b \in B$. However, this homeomorphism is not canonical. Depending on the context, we will denote a fibre bundle using only the total space E, by $\pi\colon E \to B$, or by $F \to E \to B$, where $F \to E$ refers to a homeomorphism between F and $\pi^{-1}(\{b\})$ for some $b \in B$.

Example A.2 We now give some examples of fibre bundles.

(i) A covering space is a fibre bundle where π is a local homeomorphism, and consequently F is discrete.

(ii) If μ is the Möbius band, then the projection $\pi\colon \mu \to S^1$ onto its core circle is a fibre bundle with fibre $[-1,1]$.

(iii) Let $\varphi\colon F \to F$ be a homeomorphism. Then

$$M_\varphi := I \times F/_{(1,x)\sim(0,\varphi(x))},$$

called the *mapping torus* of φ, is a fibre bundle over S^1 with fibre F, where $\pi([(t,x)]) := t$ for $t \in I$ and $x \in F$. When $F = [-1,1]$ and $\varphi(x) = -x$, we recover the previous example.

The following key result allows one to relate the homotopy groups of F, E, and B in a fibre bundle.

Theorem A.3 *Let $F \xrightarrow{i} E \xrightarrow{\pi} B$ be a fibre bundle. Choose basepoints $f \in F$, $e \in E$, and $b \in B$ such that $i(f) = e$ and $\pi(e) = b$. Then there is a long exact sequence*

$$\cdots \longrightarrow \pi_n(F,f) \xrightarrow{i_*} \pi_n(E,e) \xrightarrow{\pi_*} \pi_n(B,b) \xrightarrow{\partial} \cdots.$$

The boundary map $\partial\colon \pi_n(B,b) \to \pi_{n-1}(F,f)$ is constructed by representing a class in $\pi_n(B,b)$ by a map $u\colon (D^n, S^{n-1}) \to (B,b)$, lifting it to a map $\tilde{u}\colon (D^n, S^{n-1}) \to (E,F)$, and taking $[\tilde{u}|_{S^{n-1}}]$.

Definition A.4 Let $\pi\colon E \to B$ and $\pi'\colon E' \to B'$ be fibre bundles. A *bundle morphism* consists of continuous maps $\varepsilon\colon E \to E'$ and $\beta\colon B \to B'$ such that the following diagram commutes:

$$\begin{array}{ccc} E & \xrightarrow{\varepsilon} & E' \\ {\scriptstyle\pi}\downarrow & & \downarrow{\scriptstyle\pi'} \\ B & \xrightarrow{\beta} & B'. \end{array}$$

Bundles and morphisms between them form a category, and hence we can talk about bundle isomorphisms. If the base space B is contractible, then every bundle over it is *trivial*; that is, equivalent to a product.

Definition A.5 We say that $s\colon B \to E$ is a *section* of the bundle $\pi\colon E \to B$ if $\pi \circ s = \mathrm{Id}_B$.

Not every bundle admits a section. For example, consider the non-trivial double cover of S^1. A section gives a splitting of the homotopy long exact sequence, which provides an obstruction to finding a section.

Definition A.6 Let $\pi\colon E \to B$ be a fibre bundle with fibre F. Given a continuous map $\phi\colon B' \to B$, we can form the *pullback* bundle $\pi'\colon \phi^*E \to B'$ by setting

$$\phi^*E := \{(b', e) \in B' \times E\colon \phi(b') = \pi(e)\}$$

and $\pi'(b', e) := b'$. Then this is also a fibre bundle with fibre F. Furthermore, if we set $\varepsilon(b', e) := e$, then (ε, ϕ) is a bundle morphism; that is, the following diagram is commutative:

$$\begin{array}{ccc} \phi^*E & \xrightarrow{\varepsilon} & E \\ {\scriptstyle \pi'}\downarrow & & \downarrow{\scriptstyle \pi} \\ B' & \xrightarrow{\phi} & B. \end{array}$$

If s is a section of $\pi\colon E \to B$, then we can define its pullback as

$$\phi^* s(b') := (b', s(\phi(b'))).$$

Analogously to smooth structures on manifolds, one can define various structures on bundles using atlases. Recall that a *topological group* is a group G that is endowed with a topology such that the product $G \times G \to G$ and the inverse $G \to G$ are both continuous.

Definition A.7 Let $F \to E \xrightarrow{\pi} B$ be a fibre bundle, and let G be a topological group admitting a left homeomorphism action on F. A *G-atlas* on the bundle is a set of local trivialisations

$$\{(U_i, h_i\colon \pi^{-1}(U_i) \to U_i \times F)\colon i \in \mathcal{I}\},$$

where $\mathcal{I}$ is an index set and $\{U_i\colon i \in \mathcal{I}\}$ is an open cover of B, such that the transition maps

$$h_j \circ h_i^{-1}\colon (U_i \cap U_j) \times F \to (U_i \cap U_j) \times F$$

are of the form $h_j \circ h_i^{-1}(b, x) = (b, t_{ij}(b)x)$ for $(b, x) \in (U_i \cap U_j) \times F$ and continuous *transition functions* $t_{ij}\colon U_i \cap U_j \to G$.

Two G-atlases are *equivalent* if their union is also a G-atlas. A *G-structure* on a fibre bundle is an equivalence class of G-atlases.

The transition functions satisfy the following conditions:

(i) $t_{ii} \equiv 1$,
(ii) $t_{ij} = t_{ji}^{-1}$, and
(iii) $t_{ij}t_{jk} = t_{ik}$.

Hence, the transition functions form a Čech one-cocycle.

Closely related to the notion of G-bundles are principal bundles.

Definition A.8 Let G be topological group. Then a *principal G-bundle* is a fibre bundle $\pi\colon P \to B$ such that P admits a continuous right action of G that preserves the fibres and acts by homeomorphisms on each fibre freely and transitively.

In particular, the fibre of a principal G-bundle is homeomorphic to G, though not canonically. The homeomorphism becomes canonical once we fix the identity element of the fibre. So a principal bundle is trivial if and only if it admits a section.

Suppose that $\pi\colon P \to B$ is a principal G-bundle, and that G acts on a topological space F on the left by homeomorphisms. Then we can form the *associated bundle*

$$P \times_G F := P \times F/_{(p,f)\sim(pg,g^{-1}f)},$$

which is a fibre bundle with a G-structure.

Conversely, given a fibre bundle $\pi\colon E \to B$ with a G-structure, we can associate to it a principal G-bundle over B by using the transition functions t_{ij} to glue $U_i \times G$ and $U_j \times G$.

Theorem A.9 *If G is a compact topological group, then there is a universal principal G-bundle $EG \to BG$ such that the pullback operation induces a bijection between the set of homotopy classes $[B, BG]$ and isomorphism classes of principal G-bundles over B.*

Furthermore, the universal bundle is a universal terminal object in the category of principal G-bundles, and is hence unique up to homotopy equivalence. It is characterised by the property that EG is contractible.

To show its existence, it suffices to construct a contractible space EG that admits a free and proper G-action, and set $BG := EG/G$. Milnor gave the following construction. Let

$$E_nG := \left\{ (g_1, \ldots, g_n, t_1 \ldots, t_n) \in G^n \times I^n \colon \sum_{i=1}^{n} t_i = 1 \right\},$$

and set EG to be the direct limit of the E_nG as $n \to \infty$. One can think of elements of EG as finite convex combinations of elements of G.

Definition A.10 Given topological spaces X and Y, we define their *join* $X * Y$ as $X \times I \times Y/\sim$, where $(x, 0, y) \sim (x, 0, y')$ and $(x, 1, y) \sim (x', 1, y)$ for every x, $x' \in X$ and y, $y' \in Y$.

Intuitively, we connect each point of X and each point of Y with a unit interval. Then $E_n G$ is homeomorphic to the n-fold join $G * \cdots * G$.

Example A.11 We now compute the universal principal bundle for several important groups G.

(i) We have

$$E_n \mathbb{Z}_2 = S^0 * \cdots * S^0 \approx S^{n-1}.$$

The free $\mathbb{Z}_2$-action is the antipodal map, hence $B_n \mathbb{Z}_2 = \mathbb{RP}^{n-1}$. It follows that $E\mathbb{Z}_2 = S^\infty$ and $B\mathbb{Z}_2 = \mathbb{RP}^\infty$.

(ii) We have

$$E_n S^1 = S^1 * \cdots * S^1 = S^{2n-1} \subseteq \mathbb{C}^n.$$

The S^1-action is given by complex multiplication, hence $B_n S^1 = \mathbb{CP}^{n-1}$. So $ES^1 = S^\infty$ and $BS^1 = \mathbb{CP}^\infty$.

(iii) Let $V_k(\mathbb{R}^n)$ be the *Stiefel manifold* of k-frames in $\mathbb{R}^n$; that is, tuples

$$(v_1, \ldots, v_k) \in (\mathbb{R}^n)^k$$

that are linearly independent. We write $V_k(\mathbb{R}^\infty)$ for their direct limit as $n \to \infty$. This is a contractible space that admits a free and proper $\mathrm{GL}(k, \mathbb{R})$-action, hence $E\mathrm{GL}(k, \mathbb{R}) = V_k(\mathbb{R}^\infty)$. Then

$$B\mathrm{GL}(k, \mathbb{R}) = E\mathrm{GL}(k, \mathbb{R})/\mathrm{GL}(k, \mathbb{R}) = \mathrm{Gr}_k(\mathbb{R}^\infty),$$

the *Grassmannian* of k-planes in $\mathbb{R}^\infty$. Since $O(k)$ is a deformation retract of $\mathrm{GL}(k)$ by the Gram–Schmidt process, $BO(k) \simeq BGL(k)$.

(iv) Similarly, $EU(k) = V_k(\mathbb{C}^\infty)$ and $BU(k) = \mathrm{Gr}_k(\mathbb{C}^\infty)$, where $V_k(\mathbb{C}^\infty)$ is the Stiefel manifold of complex k-frames in $\mathbb{C}^\infty$, and $\mathrm{Gr}_k(\mathbb{C}^\infty)$ is the Grassmannian of complex k-planes in $\mathbb{C}^\infty$.

Definition A.12 Let $\mathbb{F}$ be either $\mathbb{R}$ or $\mathbb{C}$. A rank n *vector bundle* over $\mathbb{F}$ is a fibre bundle $\mathbb{F}^n \to E \xrightarrow{\pi} B$ such that

(i) $\pi^{-1}(b)$ has the structure of an $\mathbb{F}$-vector space for every $b \in B$;

(ii) every point of B has a neighbourhood U and a local trivialisation

$$h \colon \pi^{-1}(U) \to U \times \mathbb{F}^n$$

such that

$$h|_{\pi^{-1}(\{b\})}\colon \pi^{-1}(\{b\}) \to \{b\} \times \mathbb{F}^n$$

is a linear isomorphism for every $b \in U$.

Equivalently, a rank n vector bundle over $\mathbb{F} \in \{\mathbb{R}, \mathbb{C}\}$ is a fibre bundle with fibre $\mathbb{F}^n$ and a $\mathrm{GL}(n, \mathbb{F})$-structure. One can construct the associated principal $\mathrm{GL}(n, \mathbb{F})$-bundle by considering the bundle of n-frames in the fibres of the vector bundle. Hence, rank n vector bundles over $\mathbb{F}$ with base space B correspond to homotopy classes $[B, B\mathrm{GL}(n, \mathbb{F})]$. Given a continuous map

$$f\colon B \to B\mathrm{GL}(n, \mathbb{F}) = \mathrm{Gr}_n(\mathbb{F}^\infty),$$

the corresponding vector bundle over B is obtained by pulling back the *tautological bundle*

$$\gamma^n_{\mathbb{F}} := E\mathrm{GL}(n, \mathbb{F}) \times_{\mathrm{GL}(n,\mathbb{F})} \mathbb{F}^n,$$

whose fibre over $p \in \mathrm{Gr}_n(\mathbb{F}^\infty)$ is the subspace p itself.

A fundamental operation on vector bundles is the *direct sum* or *Whitney sum* operation. Given vector bundles ξ and η over the same base space B, we can define their direct sum $\xi \oplus \eta$ as a vector bundle over B whose fibre over $b \in B$ is $\xi_b \oplus \eta_b$. Its precise definition is the following:

Definition A.13 Let ξ and η be vector bundles over B. Then

$$\xi \oplus \eta := \Delta^*(\xi \times \eta),$$

where $\Delta\colon B \to B \times B$ is the diagonal map, and $\xi \times \eta$ is the product bundle over $B \times B$.

Given a real rank n vector bundle ξ, its structure group is $\mathrm{GL}(n, \mathbb{R})$. Fixing a Riemannian metric on each fibre of ξ; that is, a positive definite, symmetric bilinear form that varies continuously reduces the structure group to $\mathrm{O}(n)$. Indeed, the associated principal $\mathrm{O}(n)$-bundle consists of orthonormal n-frames in the fibres of ξ. Every vector bundle admits a Riemannian metric, and so we can reduce its structure group to $\mathrm{O}(n)$. Note that $\mathrm{O}(n)$ is a deformation retract of $\mathrm{GL}(n, \mathbb{R})$ by the Gram–Schmidt process.

Fixing a consistent orientation on the fibres of ξ reduces the structure group to $\mathrm{GL}_+(n, \mathbb{R})$. If we are also given a Riemannian metric and an orientation, the structure group is $\mathrm{SO}(n)$. Recall that $\mathrm{SO}(2) \approx S^1$ and $\pi_1(\mathrm{SO}(n)) \cong \mathbb{Z}_2$ for $n > 2$.

Definition A.14 For $n \geq 2$, the *spin group* Spin(n) is the connected double cover of SO(n). Let ξ be an oriented rank n real vector bundle over B. Choose a Riemannian metric on ξ, and write $p\colon P(\xi) \to B$ for the associated principal SO(n)-bundle of positive orthonormal n-frames in ξ. A *lift of the structure group of* ξ *to Spin*(n) is a pair

$$(\pi\colon P \to B, \epsilon\colon P \to P(\xi)),$$

where π is a principal Spin(n)-bundle and ϵ is a bundle morphism that is fibrewise the non-trivial double cover Spin(n) $\to$ SO(n), and such that $p \circ \epsilon = \pi$. We say that the lifts $(\pi\colon P \to B, \epsilon)$ and $(\pi'\colon P' \to B, \epsilon')$ are *equivalent* if there is a bundle isomorphism $\iota\colon P \to P'$ such that $\epsilon' \circ \iota = \epsilon$; that is, the following diagram is commutative:

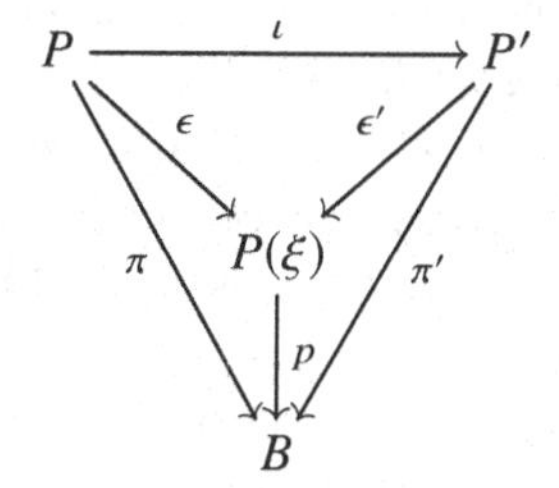

A *spin structure on* ξ is an equivalence class of lifts of the structure group of ξ to Spin(n).

A.2 Obstruction Theory

Obstruction theory deals with deciding whether a fibre bundle admits a section, and if it does, finding their homotopy classes. Suppose that $\pi\colon E \to B$ is a fibration with fibre F, and that B is an n-dimensional CW complex. We will write $\text{sk}_i(B)$ for the i-skeleton of B. If c is an i-cell with characteristic map $\chi_c\colon D^i \to \text{sk}_i(B)$, then the pullback χ_c^*E is trivial, since D^i is contractible.

We can start constructing a section s by defining it over $\text{sk}_0(B)$, and extending it recursively over $\text{sk}_i(B)$ for $i = 1, 2, \ldots, n$. Suppose we have extended s over $\text{sk}_i(B)$ and would now like to extend it to $\text{sk}_{i+1}(B)$. Given an $(i+1)$-cell c, the pullback χ_c^*E is trivial; that is, isomorphic to the product $D^{i+1} \times F$. Let us write $s_c := (\chi_c|_{S^i})^* s$, which we view as a section of $S^i \times F$. Then the obstruction to extending s over c is

$$[\pi_F \circ s_c] \in \pi_i(F),$$

where $\pi_F\colon D^{i+1} \times F \to F$ is the projection. However, note that, in general, there is no canonical identification between $\pi_i(F_b)$ and $\pi_i(F)$ for $b \in B$ and $F_b := \pi^{-1}(\{b\})$, so we will be using cohomology with twisted coefficients. The following is the fundamental result of obstruction theory.

Theorem A.15 *Let $F \to E \to B$ be a fibre bundle. Suppose that $s\colon sk_i(B) \to E$ is a section defined over the i-skeleton of B. With notation as above, the assignment $c \mapsto [\pi_F \circ s_c]$ is an $(i+1)$-cocycle in the twisted complex $C^{i+1}(B;\pi_i(F))$. Its cohomology class is zero if and only if s can be modified over $sk_i(B)$ such that it extends over $sk_{i+1}(B)$.*

If $A \subset B$ is a subcomplex and s is already defined on A, then a relative version of the theorem holds, with the cocycle lying in $C^{i+1}(B,A;\pi_i(F))$.

If we cannot extend the section to $\mathrm{sk}_{i+1}(B)$ due to the non-vanishing of the cocycle $c \mapsto [\pi_F \circ s_c]$, we call its cohomology class the *first obstruction* to extending the section.

Theorem A.15 can also be used to construct a homotopy between two sections s_0 and s_1 of the bundle $\pi\colon E \to B$, as follows. Consider the projection $\pi_B\colon B \times I \to B$. Then $s_k' := (\pi_B|_{B\times\{k\}})^* s_k$ is a section of $\pi_B^* E$ over $B \times \{k\}$ for $k \in \{0,1\}$. Then a homotopy between s_0 and s_1 is equivalent to a section of $\pi_B^* E$ extending s_0' and s_1', which is covered by the relative version of Theorem A.15.

Alternatively, we can try to construct the homotopy recursively over the skeleta. Suppose that s_0 and s_1 already agree on $\mathrm{sk}_{i-1}(B)$, and let c be an i-cell with characteristic map $\chi_c\colon D^i \to \mathrm{sk}_i(B)$. If we write

$$s_c^k := (\chi_c|_{S^{i-1}})^* s_k$$

for $k \in \{0,1\}$, then

$$s_c^0|_{S^{i-1}} = s_c^1|_{S^{i-1}}.$$

As $\chi_c^* E = D^i \times F$, the sections s_c^0 to s_c^1 can be viewed as maps $D^i \to F$, and the obstruction to finding a homotopy between them relative to S^{i-1} lies in $\pi_i(F)$. Hence, our obstruction cocycle lies in $C^i(B;\pi_i(F))$. The two approaches can be shown to be equivalent by considering the product CW decomposition of $B \times I$.

A.3 Characteristic Classes

Characteristic classes are certain natural cohomology classes in the cohomology of the base space of a fibre bundle with structure group G. By Theorem A.9, isomorphism classes of principal G-bundles over a space B are in bijection with the set of homotopy classes $[B, BG]$ via pulling back EG. Suppose that $f\colon B \to BG$ is a continuous map and $E = f^* EG$. Then a cohomology class $c \in H^*(BG; R)$ for some commutative ring R induces an element $c(E) := f^*c \in H^*(B;R)$. We call $c(E)$ a characteristic class. Characteristic classes are usually defined by finding a nice generating set of the cohomology ring $H^*(BG;R)$. We will focus on the case of real and complex vector bundles, when G is $\mathrm{GL}(n,\mathbb{R})$ or $\mathrm{GL}(n,\mathbb{C})$, and R is either $\mathbb{Z}$ or $\mathbb{Z}_2$. The characteristic classes that arise are the Stiefel–Whitney classes, the Chern classes, the Pontryagin classes, and the Euler class. These classes also have obstruction-theoretic interpretations.

A.3.1 Stiefel–Whitney classes

Let ξ be a real rank n vector bundle over the base space B. For an integer $i \geq 0$, the *ith Stiefel–Whitney class* is an element $w_i(\xi) \in H^i(B;\mathbb{Z}_2)$. We write

$$w(\xi) := w_0(\xi) + w_1(\xi) + \cdots \in H^*(B;\mathbb{Z}_2)$$

for the *total Stiefel–Whitney class* of ξ. These are completely characterised by the following axioms:

(i) We have $w_0(\xi) = 1$ and $w_i(\xi) = 0$ for $i > n$.
(ii) If γ_1^1 is the tautological line bundle over $\mathbb{RP}^1$ (whose total space is the interior of the Möbius strip), then we have the *normalisation* $w(\gamma_1^1)=1+x$, where x is the generator of $H^1(\mathbb{RP}^1;\mathbb{Z}_2) \cong \mathbb{Z}_2$.
(iii) The *Whitney sum formula* $w(\xi \oplus \eta) = w(\xi) \cup w(\eta)$ holds.
(iv) The Stiefel–Whitney class is *natural*, in the sense that if $f\colon B' \to B$ is a continuous map and ξ is a vector bundle over B, then $w(f^*\xi) = f^*w(\xi)$.

In practice, the preceding axioms suffice for any computation involving Stiefel–Whitney classes. Their existence follows from the following result; see Milnor–Stasheff [118, theorem 7.1] for a proof.

Theorem A.16 *The cohomology ring $H^*(Gr_n(\mathbb{R}^\infty);\mathbb{Z}_2)$ is a polynomial algebra over $\mathbb{Z}_2$ freely generated by the Stiefel–Whitney classes*

$$w_1(\gamma_{\mathbb{R}}^n),\ldots,w_n(\gamma_{\mathbb{R}}^n).$$

The Stiefel–Whitney classes have the following obstruction-theoretic interpretation, which was actually the original definition. Let ξ be a rank n real vector bundle over B. We write $V_k(\xi)$ for the Stiefel manifold bundle associated to ξ, which consists of k-frames in the fibres of ξ. Note that its fibre $V_k(\mathbb{R}^n)$ is $(n-k-1)$-connected, and $\pi_{n-k}(V_k(\mathbb{R}^n))$ is either $\mathbb{Z}_2$ or $\mathbb{Z}$. The first obstruction to defining a section is an element of the twisted cohomology group

$$H^{n-k+1}(B; \pi_{n-k}(V_k(\xi))).$$

Its image in $H^{n-k+1}(B; \mathbb{Z}_2)$ is precisely $w_{n-k+1}(\xi)$. If B is a closed manifold and $s_1, \dots, s_k$ are generic sections of ξ, then let

$$\Sigma := \{ b \in B : \mathrm{rk}(s_1, \dots, s_k) < k \}.$$

Then $[\Sigma]$ is Poincaré dual to $w_{n-k+1}(\xi)$.

Given a manifold M, we have $w_1(M) = 0$ if and only if M is orientable. If M is oriented, then $w_2(M) = 0$ if and only if M admits a spin structure.

A.3.2 The Euler class

An orientation of a rank n real vector bundle ξ over B is a choice of orientations of the fibres ξ_b for $b \in B$. These are coherent in the sense that there are linearly independent local sections $s_1, \dots, s_n$ in a neighbourhood of each point of B that induce the local orientations. For every $b \in B$, this gives a preferred generator of $H^n(\xi_b, \xi_b \setminus \{0\}; \mathbb{Z})$. Alternatively, it is a reduction of the structure group of ξ from $\mathrm{GL}(n, \mathbb{R})$ to $\mathrm{GL}_+(n, \mathbb{R})$. The following theorem is proven in Milnor and Stasheff [118, theorem 9.1]:

Theorem A.17 *Let ξ be an oriented rank n vector bundle with total space E and base B, and write E_0 for E minus the zero-section. Then $H^k(E, E_0) = 0$ for $k < n$, and there is a unique class $o \in H^n(E, E_0)$ that restricts to the preferred generator of $H^n(\xi_b, \xi_b \setminus \{0\})$ for every $b \in B$.*

The *Euler class* of an oriented rank n vector bundle ξ over B is an element $e(\xi) \in H^n(B; \mathbb{Z})$. We define it as the image of o in Theorem A.17 under the composition of the restriction maps

$$H^n(E, E_0) \to H^n(E) \xrightarrow{\sim} H^n(B).$$

When M is a smooth, oriented manifold, then

$$\langle e(TM), [M] \rangle = \chi(M),$$

which is why $e(\xi)$ is called Euler class. Just like the Stiefel–Whitney classes, $e(\xi)$ is natural under bundle morphisms. If $-\xi$ denotes ξ with its orientation reversed, then $e(-\xi) = -e(\xi)$. If the rank of ξ is odd, then multiplication by -1 induces an orientation-reversing automorphism of ξ, hence $e(\xi) = 0$. Compare this with the fact that $\chi(M) = 0$ for an odd-dimensional manifold. The natural homomorphism

$$H^n(B;\mathbb{Z}) \to H^n(B;\mathbb{Z}_2)$$

maps $e(\xi)$ to $w_n(\xi)$.

The Euler class satisfies the product formulas

$$e(\xi \oplus \xi') = e(\xi) \cup e(\xi') \text{ and } e(\xi \times \xi') = e(\xi) \times e(\xi').$$

However, unlike the total Stiefel–Whitney class, $e(\xi)$ is not a unit in the cohomology ring of the base, hence is not a stable invariant: If n is even, then $e(TS^n) = \mathrm{PD}(2[S^n])$, but

$$e(TS^n \oplus \varepsilon^1) = e(\varepsilon^{n+1}) = 0$$

as $\nu_{S^n \subseteq \mathbb{R}^{n+1}} = \varepsilon^1$.

We now explain the obstruction-theoretic interpretation of the Euler class. If ξ admits a nowhere zero section, then $e(\xi) = 0$. When B is a smooth manifold, then let s be a generic section of ξ. Then

$$e(\xi) = \mathrm{PD}([s^{-1}(0)]).$$

A.3.3 Chern Classes

These are the analogues of the Stiefel–Whitney classes for complex vector bundles, and live in cohomology with integer coefficients of the base. Let ξ be a rank n complex vector bundle over B. The *kth Chern class* is an element $c_k(\xi) \in H^{2k}(B)$, and the *total Chern class* is

$$c(\xi) := 1 + c_1(\xi) + \cdots + c_n(\xi) \in H^*(B).$$

The Chern classes are natural with respect to bundle morphisms. Furthermore, $c_0(\xi) = 1$ and $c_n(\xi) = e(\xi)$. They also satisfy the Whitney sum formula

$$c(\xi \oplus \eta) = c(\xi) \cup c(\eta).$$

They are normalised such that $c(\gamma_n^1) = 1 - x$, where γ_n^1 is the tautological complex line bundle over $\mathbb{CP}^n$, and $x = \mathrm{PD}[\mathbb{CP}^{n-1}]$ is the generator of

$$H^*(\mathbb{CP}^n) \cong \mathbb{Z}[x]/x^{n+1}.$$

The Chern classes are stable in the sense that $c(\xi \oplus \varepsilon^k) = c(\xi)$, where ε^k is the trivial rank k complex vector bundle over B.

The existence of the Chern classes follows from the following result; see Milnor–Stasheff [118, theorem 14.5].

Theorem A.18 *The cohomology ring $H^*(Gr_n(\mathbb{C}^\infty))$ is the polynomial ring over $\mathbb{Z}$ freely generated by $c_1(\gamma_{\mathbb{C}}^n), \ldots, c_n(\gamma_{\mathbb{C}}^n)$.*

A.3.4 Pontryagin Classes

Given a real vector bundle ξ over B, its *complexification* is $\xi \otimes \mathbb{C}$, whose fibre over $b \in B$ is $\xi_b \otimes_{\mathbb{R}} \mathbb{C}$. Its underlying real vector bundle is isomorphic to $\xi \oplus \xi$, and the complex structure is given by $J(x, y) = (-y, x)$.

The map $f(x+\mathrm{i}y) = x-\mathrm{i}y$ is an isomorphism between $\xi\otimes\mathbb{C}$ and its conjugate $\overline{\xi \otimes \mathbb{C}}$, as $f(\mathrm{i}(x + \mathrm{i}y)) = -\mathrm{i}f(x + \mathrm{i}y)$. Hence

$$c(\xi \otimes \mathbb{C}) = \sum_{k=0}^{n} c_k(\xi \otimes \mathbb{C}) =$$

$$c(\overline{\xi \otimes \mathbb{C}}) = \sum_{k=0}^{n} (-1)^k c_k(\xi \otimes \mathbb{C}).$$

It follows that $2c_k(\xi \otimes \mathbb{C}) = 0$ for k odd.

Definition A.19 The *kth Pontryagin class* of a real vector bundle ξ over B is

$$p_k(\xi) := (-1)^k c_{2k}(\xi \otimes \mathbb{C}) \in H^{2k}(B).$$

The *total Pontryagin* class is

$$p(\xi) := 1 + p_1(\xi) + \cdots + p_{\lfloor n/2 \rfloor}(\xi).$$

The Pontryagin classes are natural with respect to bundle morphisms. They are stable in the sense that $p(\xi \oplus \varepsilon^k) = p(\xi)$, where ε^k is the trivial rank k real vector bundle over B. They satisfy the Whitney sum formula modulo elements of order two; that is,

$$2\big(p(\xi \oplus \eta) - p(\xi) \cup p(\eta)\big) = 0. \tag{A.1}$$

Since $\xi \otimes \mathbb{C}$ is isomorphic to $\xi \oplus \xi$, when ξ has even rank $2m$, we have $p_m(\xi) = e(\xi)^2$.

The *oriented Grassmannian* $\widetilde{\mathrm{Gr}}_n(\mathbb{R}^\infty) = BSO(n)$ is the space of oriented n-planes in $\mathbb{R}^\infty$. It is a double cover of $\mathrm{Gr}_n(\mathbb{R}^\infty)$. The corresponding tautological bundle is $\tilde{\gamma}^n_{\mathbb{R}}$. We have the following result for its cohomology ring; see Milnor–Stasheff [118, theorem 15.9].

Theorem A.20 *Let R be an integral domain containing $1/2$. Then the cohomology ring $H^*(\widetilde{Gr}_{2m+1}(\mathbb{R}^\infty); R)$ is a polynomial ring over R generated by the Pontryagin classes $p_1(\tilde{\gamma}^{2m+1}_{\mathbb{R}}), \ldots, p_m(\tilde{\gamma}^{2m+1}_{\mathbb{R}})$. Similarly, $H^*(\widetilde{Gr}_{2m}(\mathbb{R}^\infty); R)$ is a polynomial ring over R generated by $p_1(\tilde{\gamma}^{2m}_{\mathbb{R}}), \ldots, p_{m-1}(\tilde{\gamma}^{2m}_{\mathbb{R}})$ and the Euler class $e(\tilde{\gamma}^{2m}_{\mathbb{R}})$.*

References

[1] John Frank Adams, *On the non-existence of elements of Hopf invariant one*, Ann. of Math. **72** (1960), 20–104. MR141119.

[2] John Frank Adams, *On the groups J(X). IV*, Topology **5** (1966), 21–71. MR198470.

[3] Sergei I. Adian, *The unsolvability of certain algorithmic problems in the theory of groups*, Trudy Moskov. Mat. Obsc. **6** (1957), 231–98. (Russian).

[4] Ian Agol, *The virtual Haken conjecture*, Doc. Math. **18** (2013), 1045–87. With an appendix by Agol, Daniel Groves, and Jason Manning. MR3104553.

[5] Ian Agol, *Ribbon concordance of knots is a partial order* (2022). https://arxiv.org/abs/2201.03626.

[6] Selman Akbulut, *4-manifolds*, Oxford Graduate Texts in Mathematics, vol. 25, Oxford University Press, Oxford, 2016. MR3559604.

[7] Emil Artin, *Zur Isotopie zweidimensionaler Flächen im* R_4, Abh. Math. Sem. Univ. Hamburg **4** (1925), no. 1, 174–77. MR3069446.

[8] Dennis Barden, The structure of manifolds, PhD thesis, Cambridge University (1963).

[9] Stefan Behrens, Boldizsár Kalmár, Min Hoon Kim, Mark Powell, and Arunima Ray, *The disc embedding theorem: Based on the work of Michael H. Freedman*, Oxford University Press, Oxford, 2021.

[10] Steven A. Bleiler, Craig D. Hodgson, and Jeffrey R. Weeks, *Cosmetic surgery on knots*, Proceedings of the Kirbyfest (Berkeley, CA, 1998), Geom. Topol. Monogr., vol. 2, Geom. Topol. Publ., Coventry, 1999, pp. 23–34. MR1734400.

[11] Raoul Bott and Loring W. Tu, *Differential forms in algebraic topology*, Graduate Texts in Mathematics, vol. 82, Springer-Verlag, New York, 1982. MR658304.

[12] Egbert Brieskorn, *Beispiele zur Differentialtopologie von Singularitäten*, Invent. Math. **2** (1966), 1–14. MR206972.

[13] William Browder, *The Kervaire invariant of framed manifolds and its generalization*, Ann. of Math. (2) **90** (1969), 157–86. MR251736.

[14] William Browder, *Surgery on simply-connected manifolds*, Springer-Verlag, New York, 1972. Ergebnisse der Mathematik und ihrer Grenzgebiete, Band 65. MR0358813.

[15] Gerhard Burde, Heiner Zieschang, and Michael Heusener, *Knots*, extended, De Gruyter Studies in Mathematics, vol. 5, De Gruyter, Berlin, 2014. MR3156509.

[16] Andrew J. Casson, Three-dimensional topology. Unpublished lecture notes.

[17] Andrew J. Casson and Cameron McA. Gordon, *On slice knots in dimension three*, Algebraic and geometric topology (Proc. Sympos. Pure Math., Stanford Univ., Stanford, CA, 1976), Part 2, 1978, pp. 39–53. MR520521.

[18] Andrew J. Casson and Cameron McA. Gordon, *Cobordism of classical knots*, à la recherche de la topologie perdue, 1986, pp. 181–99. With an appendix by P. M. Gilmer. Birkhäuser Verlag, Basel. MR900252.

[19] Jean Cerf, *Topologie de certains espaces deplongements*, Bull. Soc. Math. France **89** (1961), 227–380. MR140120.

[20] Jean Cerf, *Sur les diffeomorphismes de la sphère de dimension trois* ($\Gamma_4 = 0$), Lecture Notes in Mathematics, no. 53, Springer-Verlag, Berlin, 1968. MR0229250.

[21] Jean Cerf, *La stratification naturelle des espaces de fonctions differentiables réelles et le théorème de la pseudo-isotopie*, Inst. Hautes Études Sci. Publ. Math. **39** (1970), 5–173. MR292089.

[22] T A. Chapman, *Topological invariance of Whitehead torsion*, Amer. J. Math. **96** (1974), 488–97. MR391109.

[23] Tim D. Cochran, Kent E. Orr, and Peter Teichner, *Knot concordance, Whitney towers and L^2-signatures*, Ann. of Math. (2) **157** (2003), no. 2, 433–519. MR1973052.

[24] Marshall M. Cohen, *A course in simple-homotopy theory*, Springer-Verlag, New York, 1973. Graduate Texts in Mathematics, vol. 10. MR0362320.

[25] Ralph L. Cohen, *The immersion conjecture for differentiable manifolds*, Ann. of Math. (2) **122** (1985), no. 2, 237–328. MR808220.

[26] Alexander Coward and Marc Lackenby, *An upper bound on Reidemeister moves*, Amer. J. Math. **136** (2014), no. 4, 1023–66. MR3245186.

[27] Marc Culler, Nathan M. Dunfield, Matthias Goerner, and Jeffrey R. Weeks, *SnapPy, a computer program for studying the geometry and topology of 3-manifolds.* Accessed on 23 May, 2021 at http://snappy.computop.org.

[28] A. Davies, P. Veličković, L. Buesing, S. Blackwell, D. Zheng, N. Tomašev, R. Tanburn, P. Battaglia, C. Blundell, A. Juhász, M. Lackenby, G. Williamson, D. Hassabis, and P. Kohli, *Advancing mathematics by guiding human intuition with AI*, Nature **600** (2021), 70–4.

[29] Simon K. Donaldson, *An application of gauge theory to four-dimensional topology*, J. Differential Geom. **18** (1983), no. 2, 279–315. MR710056.

[30] Tobias Ekholm and András Szűcs, *The group of immersions of homotopy $(4k - 1)$-spheres*, Bull. London Math. Soc. **38** (2006), no. 1, 163–76. MR2201615.

[31] Yakov Eliashberg, *Classification of overtwisted contact structures on 3-manifolds*, Invent. Math. **98** (1989), no. 3, 623–37. MR1022310.

[32] John Etnyre, *Cobordisms and Morse theory*, 2019. https://etnyre.math.gatech .edu/professionalstuff/morse.html.

[33] Ronald Fintushel and Ronald Stern, *Knots, links, and 4-manifolds*, Invent. Math. **134** (1998), 363–400.

[34] Ralph H. Fox, *A quick trip through knot theory*, Topology of 3-manifolds and related topics (Proc. the Univ. of Georgia Inst., 1961), Prentice-Hall, Englewood Cliffs, NJ, 1962, pp. 120–67. MR0140099.

[35] Ralph H. Fox, *Rolling*, Bull. Amer. Math. Soc. **72** (1966), 162–4. MR184221.
[36] Ralph H. Fox and John W. Milnor, *Singularities of* 2*-spheres in* 4*-space and cobordism of knots*, Osaka Math. J. **3** (1966), 257–67. MR211392.
[37] Michael H. Freedman, *The topology of four-dimensional manifolds*, J. Differential Geom. **17** (1982), no. 3, 357–453. MR679066.
[38] Michael H. Freedman and Frank Quinn, *Topology of 4-manifolds*, Princeton Mathematical Series, vol. 39, Princeton University Press, Princeton, NJ, 1990. MR1201584.
[39] Peter Freyd, David Yetter, Jim Hoste, William B. R. Lickorish, Kenneth Millett, and Adria Ocneanu, *A new polynomial invariant of knots and links*, Bull. Amer. Math. Soc. (N.S.) **12** (1985), no. 2, 239–46. MR776477.
[40] Kenji Fukaya, Yong-Geun Oh, Hiroshi Ohta, and Kaoru Ono, *Lagrangian intersection Floer theory: Anomaly and obstruction. Part I*, AMS/IP Studies in Advanced Mathematics, vol. 46, American Mathematical Society, Providence, RI/International Press, Somerville, MA, 2009. MR2553465.
[41] Kenji Fukaya, Yong-Geun Oh, Hiroshi Ohta, and Kaoru Ono, *Lagrangian intersection Floer theory: Anomaly and obstruction. Part II*, AMS/IP Studies in Advanced Mathematics, vol. 46, American Mathematical Society, Providence, RI/International Press, Somerville, MA, 2009. MR2548482.
[42] Dimitry B. Fuks and Vladimir A. Rokhlin, *Beginner's course in topology: Geometric chapters* (Universitext), Springer-Verlag, Berlin, 1984. Translated from the Russian by A. Iacob, Springer Series in Soviet Mathematics. MR759162.
[43] Mikio Furuta, *Monopole equation and the* $\frac{11}{8}$*-conjecture*, Math. Res. Lett. **8** (2001), no. 3, 279–91. MR1839478.
[44] David Gabai, *Foliations and the topology of 3-manifolds*, J. Differential Geom. **18** (1983), 445–503.
[45] Murray Gerstenhaber and Oscar S. Rothaus, *The solution of sets of equations in groups*, Proc. Natl. Acad. Sci. U.S.A. **48** (1962), 1531–33. MR166296.
[46] Paolo Ghiggini, *Knot Floer homology detects genus-one fibred knots*, Amer. J. Math. **130** (2008), no. 5, 1151–1169.
[47] Emmanuel Giroux, *Géométrie de contact: de la dimension trois vers les dimensions supérieures*, Proceedings of the International Congress of Mathematicians, vol. 2 (Beijing, 2002), pp. 405–14. MR1957051.
[48] Robert E. Gompf and András I. Stipsicz, 4*-manifolds and Kirby calculus*, Graduate Studies in Mathematics, vol. 20, American Mathematical Society, Providence, RI, 1999. MR1707327.
[49] Cameron McA. Gordon, *Ribbon concordance of knots in the* 3*-sphere*, Math. Ann. **257** (1981), no. 2, 157–170. MR634459.
[50] Cameron McA. Gordon and Richard A. Litherland, *On the signature of a link*, Invent. Math. **47** (1978), no. 1, 53–69. MR500905.
[51] Cameron McA. Gordon and John Luecke, *Knots are determined by their complements*, J. Amer. Math. Soc. **2** (1989), no. 2, 371–415. MR965210.
[52] Joshua Evan Greene, *The lens space realization problem*, Ann. of Math. (2) **177** (2013), no. 2, 449–511. MR3010805.
[53] Joshua Evan Greene, *Alternating links and definite surfaces*, Duke Math. J. **166** (2017), no. 11, 2133–51. With an appendix by András Juhász and Marc Lackenby. MR3694566.

[54] Wolfgang Haken, *Theorie der Normalflächen*, Acta Math. **105** (1961), 245–375. MR141106.

[55] Wolfgang Haken, *Über das Homöomorphieproblem der 3-Mannigfaltigkeiten. I*, Math. Z. **80** (1962), 89–120. MR160196.

[56] Andrew J. S. Hamilton, *The triangulation of* 3*-manifolds*, Quart. J. Math. Oxford Ser. (2) **27** (1976), no. 105, 63–70. MR407848.

[57] Allen E. Hatcher, *Algebraic topology*, Cambridge University Press, Cambridge, 2002.

[58] Allen E. Hatcher, *Notes on basic 3-manifold topology*, 2007. https://pi.math .cornell.edu/ hatcher/3M/3Mfds.pdf.

[59] Allen E. Hatcher, *The Kirby torus trick for surfaces*, arXiv:1312.3518 (2022).

[60] Matthew Hedden, *Knot Floer homology and Whitehead doubles*, Geom. Topol. **11** (2007), 2277–338.

[61] Geoffrey Hemion, *On the classification of homeomorphisms of 2-manifolds and the classification of 3-manifolds*, Acta Math. **142** (1979), no. 1–2, 123–55. MR512214.

[62] John Hempel, *3-manifolds*, Princeton University Press, Princeton, NJ/University of Tokyo Press, Tokyo, 1976. Ann. of Math. Studies, vol. 86. MR0415619.

[63] Michael A. Hill, Michael J. Hopkins, and Douglas C. Ravenel, *On the nonexistence of elements of Kervaire invariant one*, Ann. of Math. (2) **184** (2016), no. 1, 1–262. MR3505179.

[64] Mikami Hirasawa and Makoto Sakuma, *Minimal genus Seifert surfaces for alternating links*, KNOTS '96 (Tokyo), 1997, pp. 383–94. MR1664976.

[65] Morris W. Hirsch, *Differential topology*, Graduate Texts in Mathematics, no. 33, Springer-Verlag, New York, 1976. MR0448362.

[66] Morris W. Hirsch and Barry Mazur, *Smoothings of piecewise linear manifolds*, Princeton University Press, Princeton, NJ/University of Tokyo Press, Tokyo, 1974. Ann. Math. Stud., no. 80. MR0415630.

[67] Ko Honda, William Kazez, and Gordana Matić, *On the contact class in Heegaard Floer homology*, J. Differential Geom. **83** (2009), no. 2, 289–311.

[68] William H. Jaco and Peter B. Shalen, *Seifert fibered spaces in 3-manifolds*, Mem. Amer. Math. Soc. **21** (1979), no. 220, viii+192. MR539411.

[69] Bo Ju Jiang, *A simple proof that the concordance group of algebraically slice knots is infinitely generated*, Proc. Amer. Math. Soc. **83** (1981), no. 1, 189–92. MR620010.

[70] Klaus Johannson, *Homotopy equivalences of* 3*-manifolds with boundaries*, Lect. Notes Math., vol. 761, Springer, Berlin, 1979. MR551744.

[71] Vaughan F. R. Jones, *A polynomial invariant for knots via von Neumann algebras*, Bull. Amer. Math. Soc. (N.S.) **12** (1985), no. 1, 103–11. MR766964.

[72] András Juhász, *Holomorphic discs and sutured manifolds*, Algebr. Geom. Topol. **6** (2006), 1429–57.

[73] András Juhász, *Floer homology and surface decompositions*, Geom. Topol. **12** (2008), 299–350.

[74] András Juhász, Dylan Thurston, and Ian Zemke, *Naturality and mapping class groups in Heegard Floer homology*, Mem. Amer. Math. Soc. **273** (2021), no. 1338, v+174. MR4337438.

[75] Jeremy Kahn and Vladimir Marković, *Counting essential surfaces in a closed hyperbolic three-manifold*, Geom. Topol. **16** (2012), no. 1, 601–24. MR2916295.

[76] Jeremy Kahn and Vladimir Marković, *Immersing almost geodesic surfaces in a closed hyperbolic three manifold*, Ann. of Math. (2) **175** (2012), no. 3, 1127–90. MR2912704.

[77] Louis H. Kauffman, *State models and the Jones polynomial*, Topology **26** (1987), no. 3, 395–407. MR899057.

[78] Louis H. Kauffman, *State models and the Jones polynomial*, Topology **26** (1987), no. 3, 395–407. MR899057.

[79] Michel A. Kervaire, *On the Pontryagin classes of certain* $SO(n)$*-bundles over manifolds*, Amer. J. Math. **80** (1958), 632–38. MR102806.

[80] Michel A. Kervaire, *A manifold which does not admit any differentiable structure*, Comment. Math. Helv. **34** (1960), 257–70. MR139172.

[81] Michel A. Kervaire, *Le théorème de Barden-Mazur-Stallings*, Comment. Math. Helv. **40** (1965), 31–42. MR189048.

[82] Michel A. Kervaire and John W. Milnor, *Groups of homotopy spheres. I*, Ann. of Math. (2) **77** (1963), 504–37. MR148075.

[83] Robion C. Kirby, *A calculus for framed links in* S^3, Invent. Math. **45** (1978), 35–56.

[84] Robion C. Kirby, *A calculus for framed links in* S^3, Invent. Math. **45** (1978), no. 1, 35–56. MR467753.

[85] Robion C. Kirby, *The topology of* 4*-manifolds*, Lect. Notes Math., vol. 1374, Springer, Berlin, 1989. MR1001966.

[86] Robion C. Kirby, *Problems in low-dimensional topology*, 2021. www.math.berkeley.edu/~kirby/problems.ps.gz.

[87] Robion C. Kirby and Laurence C. Siebenmann, *Foundational essays on topological manifolds, smoothings, and triangulations*, Princeton University Press, Princeton, NJ/University of Tokyo Press, Tokyo, 1977. With notes by John Milnor and Michael Atiyah, Ann. Math. Stud., no. 88. MR0645390.

[88] Hellmuth Kneser, *Geschlossen Flächen in dreidimensionalen Mannigfaltigkeiten*, Jahresbericht der Deutschen Mathematiker Vereinigung **38** (1929), 248–60.

[89] Antoni A. Kosinski, *Differential manifolds*, Pure and Applied Mathematics, vol. 138, Academic Press, Inc., Boston, MA, 1993. MR1190010.

[90] Peter B. Kronheimer and Tomasz S. Mrowka, *The genus of embedded surfaces in the projective plane*, Math. Res. Lett. **1** (1994), no. 6, 797–808. MR1306022.

[91] Peter B. Kronheimer and Tomasz S. Mrowka, *Khovanov homology is an unknot-detector*, Publ. Math. Inst. Hautes Etudes Sci. **113** (2011), 97–208. MR2805599.

[92] Marc Lackenby, *Word hyperbolic Dehn surgery*, Invent. Math. **140** (2000), no. 2, 243–82. MR1756996.

[93] Marc Lackenby, *A polynomial upper bound on Reidemeister moves*, Ann. of Math. (2) **182** (2015), no. 2, 491–564. MR3418524.

[94] Marc Lackenby and Robert Meyerhoff, *The maximal number of exceptional Dehn surgeries*, Invent. Math. **191** (2013), no. 2, 341–82. MR3010379.

[95] Jerome Levine, *Invariants of knot cobordism*, Invent. Math. **8** (1969), 98–110; addendum, ibid. 8 (1969), 355. MR253348.

[96] Jerome Levine, *Knot cobordism groups in codimension two*, Comment. Math. Helv. **44** (1969), 229–44. MR246314.

[97] William B. R. Lickorish, *A representation of orientable combinatorial 3-manifolds*, Ann. of Math. (2) **76** (1962), 531–40. MR151948.

[98] William B. R. Lickorish, *An introduction to knot theory*, Graduate Texts in Mathematics, vol. 175, Springer, New York, 1997. MR1472978.

[99] William B. R. Lickorish and Kenneth C. Millett, *The new polynomial invariants of knots and links*, Math. Mag. **61** (1988), no. 1, 3–23. MR934822.

[100] Robert Lipshitz, *A cylindrical reformulation of Heegaard Floer homology*, Geom. Topol. **10** (2006), 955–1097.

[101] Richard A. Litherland, *Deforming twist-spun knots*, Trans. Amer. Math. Soc. **250** (1979), 311–31. MR530058.

[102] Charles Livingston, *Order 2 algebraically slice knots*, Proceedings of the Kirbyfest (Berkeley, CA, 1998), Geom. Topol. Monogr., vol. 2, Geom. Topol. Publ., Coventry, 1999, pp. 335–42. MR1734416.

[103] Charles Livingston, *A survey of classical knot concordance*, Handbook of knot theory, 2005, pp. 319–347. *arXiv:math/0307077v4 [math.GT]*. MR2179265.

[104] Ciprian Manolescu, *Lectures on the triangulation conjecture*, Proceedings of the Gökova Geometry-Topology Conference 2015, International Press, Somerville, MA, 2016, pp. 1–38. MR3526837.

[105] John N. Mather, *Stability of C^∞ mappings. I. The division theorem*, Ann. of Math. (2) **87** (1968), 89–104. MR232401.

[106] John N. Mather, *Stability of C^∞ mappings. III. Finitely determined map germs*, Inst. Hautes Études Sci. Publ. Math. **35** (1968), 279–308. MR275459.

[107] John N. Mather, *Stability of C^∞ mappings. II. Infinitesimal stability implies stability*, Ann. of Math. (2) **89** (1969), 254–91. MR259953.

[108] John N. Mather, *Stability of C^∞ mappings. IV. Classification of stable germs by R-algebras*, Publ. Math. Inst. Hautes Études Sci. **37** (1969), 223–48. MR275460.

[109] Barry Mazur, *Relative neighborhoods and the theorems of Smale*, Ann. of Math. (2) **77** (1963), 232–49. MR150786.

[110] William W. Menasco and Morwen B. Thistlethwaite, *The Tait flyping conjecture*, Bull. Amer. Math. Soc. (N.S.) **25** (1991), no. 2, 403–12. MR1098346.

[111] John Milnor, *On manifolds homeomorphic to the 7-sphere*, Ann. of Math. (2) **64** (1956), 399–405. MR82103.

[112] John Milnor, *Singular points of complex hypersurfaces*, Ann. Math. Stud., no. 61, Princeton University Press, Princeton, NJ/University of Tokyo Press, Tokyo, 1968. MR0239612.

[113] John W. Milnor, *A unique decomposition theorem for 3-manifolds*, Amer. J. Math. **84** (1962), 1–7. MR142125.

[114] John W. Milnor, *Morse theory*, Based on lecture notes by M. Spivak and R. Wells. Ann. Math. Stud., no. 51, Princeton University Press, Princeton, NJ, 1963. MR0163331.

[115] John W. Milnor, *Lectures on the h-cobordism theorem*, Notes by L. Siebenmann and J. Sondow, Princeton University Press, Princeton, NJ, 1965. MR0190942.

[116] John W. Milnor, *Topology from the differentiable viewpoint*, The University Press of Virginia, Charlottesville, VA, 1965. Based on notes by David W. Weaver. MR0226651.

[117] John W. Milnor, *On the 3-dimensional Brieskorn manifolds $M(p,q,r)$*, Knots, groups, and 3-manifolds (Papers dedicated to the memory of R. H. Fox), 1975, pp. 175–225. MR0418127.

[118] John W. Milnor and James D. Stasheff, *Characteristic classes*, Princeton University Press, Princeton, NJ/University of Tokyo Press, Tokyo, 1974. Ann. Math. Stud., no. 76. MR0440554.

[119] Edwin E. Moise, *Affine structures in 3-manifolds. V. The triangulation theorem and Hauptvermutung*, Ann. of Math. (2) **56** (1952), 96–114. MR48805.

[120] Edwin E. Moise, *Geometric topology in dimensions 2 and 3*, Springer, New York, 1977. Graduate Texts in Mathematics, vol. 47. MR0488059.

[121] John W. Morgan, *The Seiberg-Witten equations and applications to the topology of smooth four-manifolds*, Math. Notes, vol. 44, Princeton University Press, Princeton, NJ, 1996. MR1367507.

[122] John W. Morgan and Gang Tian, *Ricci flow and the Poincaré conjecture*, Clay Mathematics Monographs, vol. 3, American Mathematical Society, Providence, RI/Clay Mathematics Institute, Cambridge, MA, 2007. MR2334563.

[123] George D. Mostow, *Strong rigidity of locally symmetric spaces*, Princeton University Press, Princeton, NJ/University of Tokyo Press, Tokyo, 1973. Ann. Math. Stud., no. 78. MR0385004.

[124] James R. Munkres, *Elementary differential topology: Lectures given at Massachusetts Institute of Technology, Fall, 1961*, Ann. Math. Stud., no. 54 Princeton University Press, Princeton, NJ, 1966. MR0198479.

[125] James R. Munkres, *Topology*, Prentice Hall, Inc., Upper Saddle River, NJ, 2000. Second edition of [MR0464128]. MR3728284.

[126] Kunio Murasugi, *On a certain numerical invariant of link types*, Trans. Amer. Math. Soc. **117** (1965), 387–422. MR171275.

[127] Kunio Murasugi, *Jones polynomials and classical conjectures in knot theory*, Topology **26** (1987), no. 2, 187–94. MR895570.

[128] Kunio Murasugi, *On the braid index of alternating links*, Trans. Amer. Math. Soc. **326** (1991), no. 1, 237–60. MR1000333.

[129] Yi Ni, *Knot Floer homology detects fibred knots*, Invent. Math. **170** (2007), no. 3, 577–608. MR2357503.

[130] Yi Ni, *Erratum: Knot Floer homology detects fibred knots [MR2357503]*, Invent. Math. **177** (2009), no. 1, 235–8. MR2507641.

[131] Jakob Nielsen, *Untersuchungen zur Topologie der geschlossenen zweiseitigen Flächen*, Acta Math. **50** (1927), no. 1, 189–358. MR1555256.

[132] Sergei P. Novikov, *Topological invariance of rational classes of Pontrjagin*, Dokl. Akad. Nauk SSSR **163** (1965), 298–300. MR0193644.

[133] Burak Ozbagci and András I. Stipsicz, *Surgery on contact 3-manifolds and Stein surfaces*, Bolyai Society Mathematical Studies, vol. 13, Springer-Verlag, Berlin; János Bolyai Mathematical Society, Budapest, 2004. MR2114165.

[134] Peter Ozsváth, András Stipsicz, and Zoltán Szabó, *Grid homology for knots and links*, Mathematical Surveys and Monographs, vol. 208, American Mathematical Society, 2015.

[135] Peter Ozsváth and Zoltán Szabó, *Holomorphic disks and 3-manifold invariants: properties and applications*, Ann. of Math. **159** (2004), no. 3, 1159–245.

[136] Peter Ozsváth and Zoltán Szabó, *Holomorphic disks and genus bounds*, Geometry and Topology **8** (2004), 311–34.

[137] Peter Ozsváth and Zoltán Szabó, *Holomorphic disks and knot invariants*, Adv. Math. **186** (2004), no. 1, 58–116.

[138] Peter Ozsváth and Zoltán Szabó, *Holomorphic disks and topological invariants for closed three-manifolds*, Ann. of Math. **159** (2004), no. 3, 1027–158.

[139] Peter Ozsváth and Zoltán Szabó, *Heegaard Floer homology and contact structures*, Duke Math. J. **129** (2005), 39–61.

[140] Peter Ozsváth and Zoltán Szabó, *Holomorphic triangles and invariants for smooth four-manifolds*, Adv. Math. **202** (2006), 326–400.

[141] Peter Ozsváth and Zoltán Szabó, *Holomorphic disks, link invariants, and the multi-variable Alexander polynomial*, Algebr. Geom. Topol. **8** (2008), no. 2, 615–92.

[142] Peter Ozsváth and Zoltán Szabó, *Kauffman states, bordered algebras, and a bigraded knot invariant*, Adv. Math. **328** (2018), 1088–198.

[143] Richard S. Palais, *Local triviality of the restriction map for embeddings*, Comment. Math. Helv. **34** (1960), 305–12. MR123338.

[144] Christos D. Papakyriakopoulos, *On Dehn's lemma and the asphericity of knots*, Ann. of Math. (2) **66** (1957), 1–26. MR90053.

[145] Frédéric Pham, *Formules de Picard-Lefschetz généralisées et ramification des intégrales*, Bull. Soc. Math. France **93** (1965), 333–67. MR195868.

[146] Gopal Prasad, *Strong rigidity of $\mathbf{Q}$-rank 1 lattices*, Invent. Math. **21** (1973), 255–86. MR385005.

[147] Józef H. Przytycki and Pawel Traczyk, *Invariants of links of Conway type*, Kobe J. Math. **4** (1988), no. 2, 115–39. MR945888.

[148] Jessica S. Purcell, *Hyperbolic knot theory*, Graduate Studies in Mathematics, vol. 209, American Mathematical Society, Providence, RI, [2020] ©2020. MR4249621.

[149] Michael O. Rabin, *Recursive unsolvability of group theoretic problems*, Ann. of Math. (2) **67** (1958), 172–94. MR110743.

[150] Tibor Radó, *Über den Begriff der Riemannschen Flache*, Acta Sci. Math. Szeged. **2** (1925), 101–21.

[151] Jacob A. Rasmussen, *Floer homology and knot complements*, PhD thesis, Harvard University (2003).

[152] Kurt Reidemeister, *Elementare Begründung der Knotentheorie*, Abh. Math. Sem. Univ. Hamburg **5** (1927), no. 1, 24–32. MR3069462.

[153] Kurt Reidemeister, *Homotopieringe und Linsenraume*, Abh. Math. Sem. Univ. Hamburg **11** (1935), no. 1, 102–9. MR3069647.

[154] Vladimir A. Rohlin, *New results in the theory of four-dimensional manifolds*, Doklady Akad. Nauk SSSR (N.S.) **84** (1952), 221–4. MR0052101.

[155] Dale Rolfsen, *Knots and links*, Mathematics Lecture Series, vol. 7, Publish or Perish, Inc., Houston, TX, 1990. Corrected reprint of the 1976 original. MR1277811.

[156] Sucharit Sarkar and Jiajun Wang, *An algorithm for computing some Heegaard Floer homologies*, Ann. of Math. **171** (2010), no. 2, 1213–36.

[157] Horst Schubert, *Knoten und Vollringe*, Acta Math. **90** (1953), 131–286. MR72482.

[158] Matthias Schwarz, *Morse homology*, Progress in Mathematics, vol. 111, Birkhäuser Verlag, Basel, 1993. MR1239174.

[159] Alexandru Scorpan, *The wild world of 4-manifolds*, American Mathematical Society, Providence, RI, 2005. MR2136212.

[160] Peter Scott, *The geometries of 3-manifolds*, Bull. London Math. Soc. **15** (1983), no. 5, 401–87. MR705527.

[161] Paul Seidel, *Fukaya categories and Picard-Lefschetz theory*, Zurich Lectures in Advanced Mathematics, European Mathematical Society (EMS), Zürich, 2008. MR2441780.

[162] Jean-Pierre Serre, *Groupes d'homotopie et classes de groupes abeliens*, Ann. of Math. (2) **58** (1953), 258–94. MR59548.

[163] Stephen Smale, *Generalized Poincaré's conjecture in dimensions greater than four*, Ann. of Math. (2) **74** (1961), 391–406. MR137124.

[164] John R. Stallings, *On infinite processes leading to differentiability in the complement of a point*, Differential and Combinatorial Topology (A Symposium in Honor of Marston Morse), 1965, pp. 245–54. MR0180983.

[165] Norman Steenrod, *The topology of fibre bundles*, Princeton Landmarks in Mathematics, Princeton University Press, Princeton, NJ, 1999. Reprint of the 1957 edition, Princeton Paperbacks. MR1688579.

[166] Clifford H. Taubes, *The Seiberg–Witten invariants and symplectic forms*, Math. Res. Lett. **1** (1994), 809–22.

[167] Morwen B. Thistlethwaite, *A spanning tree expansion of the Jones polynomial*, Topology **26** (1987), no. 3, 297–309. MR899051.

[168] René Thom, *Quelques propriétés globales des variétés différentiables*, Comment. Math. Helv. **28** (1954), 17–86. MR61823.

[169] René Thom, *Les singularités des applications différentiables*, Ann. Inst. Fourier (Grenoble) **6** (1955/56), 43–87. MR87149.

[170] William P. Thurston, *Three-dimensional manifolds, Kleinian groups and hyperbolic geometry*, Bull. Amer. Math. Soc. (N.S.) **6** (1982), no. 3, 357–81. MR648524.

[171] William P. Thurston, *Three-dimensional geometry and topology. Vol. 1*, Princeton Mathematical Series, vol. 35, Princeton University Press, Princeton, NJ, 1997. Edited by Silvio Levy. MR1435975.

[172] Bruce Trace, *On the Reidemeister moves of a classical knot*, Proc. Amer. Math. Soc. **89** (1983), no. 4, 722–4. MR719004.

[173] Vladimir Turaev, *Introduction to combinatorial torsions*, Lectures in Mathematics ETH Zürich, Birkhauser Verlag, Basel, 2001. Notes taken by Felix Schlenk. MR1809561.

[174] Friedhelm Waldhausen, *Heegaard-Zerlegungender 3-Sphare*, Topology **7** (1968), 195–203. MR227992.

[175] Friedhelm Waldhausen, *On irreducible 3-manifolds which are sufficiently large*, Ann. of Math. (2) **87** (1968), 56–88. MR224099.

[176] Andrew H. Wallace, *Modifications and cobounding manifolds*, Canad. J. Math. **12** (1960), 503–28.

[177] George W. Whitehead, *On the homotopy groups of spheres and rotation groups*, Ann. of Math. (2) **43** (1942), 634–40. MR7107.

[178] Hassler Whitney, *The sections of a smooth n-manifold in 2n-space*, Ann. of Math. **45** (1944), 220–46.

[179] Erik Christopher Zeeman, *Twisting spun knots*, Trans. Amer. Math. Soc. **115** (1965), 471–95. MR195085.

[180] Ian Zemke, *Knot Floer homology obstructs ribbon concordance*, Ann. of Math. (2) **190** (2019), no. 3, 931–47. MR4024565.

Index

Printed in the United States
by Baker & Taylor Publisher Services